Security Awareness

Applying Practical Cybersecurity in Your World

Sixth Edition

Security Awareness

Applying Practical Cybersecurity in Your World

Mark Ciampa, Ph.D.

Cengage

Australia • Brazil • Canada • Mexico • Singapore • United Kingdom • United States

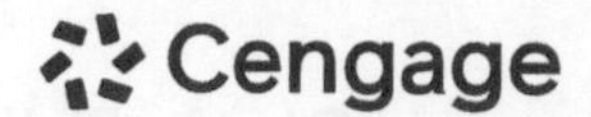

***Security Awareness: Applying Practical Cybersecurity in Your World,* Sixth Edition**
Mark Ciampa

SVP, Cengage Academic Product Management: Erin Joyner

VP, Product Management, Learning Experiences: Thais Alencar

Portfolio Product Director: Mark Santee

Portfolio Product Manager: Natalie Onderdonk

Vendor Content Manager: Manoj Kumar, Lumina Datamatics Limited

Product Assistant: Ethan Wheel

Digital Project Manager: Jim Vaughey

Marketing Manager: Mackenzie Paine

Content Acquisition Analyst: Ann Hoffman

Content Acquisition Project Manager: Haneef Abrar, Lumina Datamatics Limited

Production Service: Lumina Datamatics Limited

Designer: Erin Griffin

Cover Image Source: phochi / Shutterstock.com

Library of Congress Control Number: 2022914850

ISBN: 978-0-357-88376-1

Cengage
200 Pier 4 Boulevard
Boston, MA 02210
USA

Cengage is a leading provider of customized learning solutions with employees residing in nearly 40 different countries and sales in more than 125 countries around the world. Find your local representative at **www.cengage.com.**

To learn more about Cengage platforms and services, register or access your online learning solution, or purchase materials for your course, visit **www.cengage.com.**

Printed at CLDPC, USA, 10-22

Brief Contents

Preface **ix**

Module 1
Introduction to Cybersecurity 1

Module 2
Personal Cybersecurity 27

Module 3
Computer Security 55

Module 4
Internet Security 87

Module 5
Mobile Security 115

Module 6
Privacy 145

Appendix A
Two Rights and a Wrong: Answers 173

Glossary **177**
Index **183**

Table of Contents

Preface ix

Module 1

Introduction to Cybersecurity 1

Difficulties in Preventing Attacks 2

What Is Cybersecurity? 5
- Understanding Security 5
- Defining Cybersecurity 6
- Cybersecurity Terminology 8
- Why Cybersecurity Is Important 9

Who Are the Attackers? 13
- Cybercriminals 13
- Script Kiddies 14
- Brokers 14
- Insiders 15
- Cyberterrorists 15
- Hactivists 15
- State Actors 16

Building a Comprehensive Security Strategy 16
- Block Attacks 17
- Update Defenses 17
- Minimize Losses 17
- Use Layers 17
- Stay Alert 18

Module Summary 18

Key Terms 19

Review Questions 19

References 25

Module 2

Personal Cybersecurity 27

Personal Security Attacks 28
- Password Attacks 29
- Attacks Using Social Engineering 34
- Social Networking Risks 37
- Identity Theft 38

Creating a Defensive Stance 39

Personal Security Defenses 40
- Password Defenses 40
- Recognizing Social Engineering Attacks 43
- Avoiding Identity Theft 44
- Reducing Social Networking Risk 45

Module Summary 46

Key Terms 47

Review Questions 47

References 54

Module 3

Computer Security 55

Malware Attacks 56
- Kidnap 57
- Eavesdrop 60
- Masquerade 62
- Launchpad 64
- Sidestep 68

Computer Defenses 69
- Managing Patches 69
- Running Antimalware Software 71
- Examining Firewalls 72
- Stopping Ransomware 74

Module Summary 77

Key Terms 78

Review Questions 78

References 86

Module 4

Internet Security 87

How the Internet Works 89
- The World Wide Web 89
- Email 90

Internet Security Risks 92
- User Device Threats 92
- Threats from Web Servers 97
- Transmission Risks 99

Internet Defenses 101
- Securing the Web Browser 101
- Email Defenses 104

Module Summary 106
Key Terms 107
Review Questions 108
References 113

Module 5

Mobile Security 115

Mobile Attacks 116
- Attacks on Wireless Networks 116
- Attacks on Mobile Devices 122

Mobile Defenses 128
- Wireless Network Security 128
- Mobile Device Security 132

Module Summary 135
Key Terms 137
Review Questions 137
References 144

Module 6

Privacy 145

Data Theft 146
- What Is Being Stolen and How? 146
- Who Are the Data Thieves? 149
- What Are the Risks? 152

Privacy Protections 154
- Use Cryptography 154
- Limit Cookies 162
- Disable and Monitor MAIDs 163
- Follow Privacy Best Practices 163
- Responsibilities of Organizations 165

Module Summary 166
Key Terms 167
Review Questions 167
References 172

Appendix A

Two Rights and a Wrong: Answers 173

Glossary 177

Index 183

Preface

Cybersecurity is a major concern today for all users and businesses. This is because the volume of attacks against our devices has reached epidemic proportions. Over 450,000 new malicious programs and potentially unwanted applications are identified *each day*. The total number of instances of malware has grown from 182 million in 2013 to over 1.34 billion today.[1] Computers with limited cyber defenses that serve as "bait" to attackers will start to receive attempted logins just *52 seconds* after the device is connected to the Internet, and subsequent login attempts are then made every *15 seconds*. In one month, over *5 million attacks* are attempted on these unprotected computers.[2] Cybercrime has been called the "greatest transfer of economic wealth in history." It is estimated that cybercrime could reach $10.5 *trillion annually* by 2025.[3]

But it is not just large multinational corporations that are the targets of attackers. Anyone who uses a computing device—like a laptop, tablet, or smartphone—or who has *any* device that is connected to the Internet—like a security camera, doorbell, or thermostat—is vulnerable to attacks.

However, users still struggle with how to make their devices secure. Ask yourself this question: If you were warned that a vicious new attack was to begin within the next 15 minutes, would you know the steps to take to check the security on your devices, and then quickly fix anything that was deficient? Only a small handful of users could truthfully answer *Yes* to that question.

It is important for all computer users today to be aware of cybersecurity. Users also need to be able to take the necessary steps to defend themselves and their technology against attacks. Applying practical cybersecurity in your world has never been more important than it is right now.

Designed to provide you with the knowledge and skills needed to protect your technology devices from increasingly sophisticated cyberattacks, *Security Awareness: Applying Practical Cybersecurity in Your World, Sixth Edition* presents straightforward information to practical questions about cybersecurity defenses. *Do I need antivirus software? How can I manage all my passwords? Can I prevent a ransomware attack? Is there a way to make my web browser secure? What are patches? How can I protect my privacy?* This text is designed to help you understand the answers to these questions and many more. It also gives you the practical skills needed to protect your devices. Hands-on projects and case projects in each module allow you to apply what you have learned. *Security Awareness: Applying Practical Cybersecurity in Your World, Sixth Edition* helps you keep your technology devices secure.

Intended Audience

This book is intended to help students and professionals who need to protect their technology devices from attacks. Basic working knowledge of computers is all that is required to use this book. The book's pedagogical features are designed to provide a truly interactive learning experience to help prepare you for the challenges of securing your technology. In addition to the information presented in the text, each module includes hands-on projects that guide you through implementing practical cybersecurity in a step-by-step fashion. Each module also contains case studies, allowing you to investigate in more depth how cybersecurity impacts our lives today.

Module Descriptions

Here is a summary of the topics covered in each module of this book:

Module 1, "Introduction to Cybersecurity," introduces you to the topic of cybersecurity. It begins by examining why it is so difficult to protect our devices. It then describes cybersecurity in more detail and explores why it is important. The module also looks at who is behind attacks and then concludes by examining the steps necessary for building a comprehensive cybersecurity strategy.

Module 2, "Personal Cybersecurity," looks at attacks that are not directed at specific types of technology, like a laptop computer or a smartphone, but are broader in scope and can apply across multiple devices and technologies. These personal cybersecurity attacks are directed more towards the person rather than a specific device. This module explores personal security attacks on passwords, attacks that use social engineering, the risks of social networking, and identity theft. It also examines defenses for these attacks.

Module 3, "Computer Security," discusses how protecting your personal computing device—be it a desktop, laptop, or tablet—is a challenge, even for the most advanced computer users. This is because many different types of attacks are launched against these devices, and attackers are constantly modifying these attacks as well as creating new ones. This module investigates computer security by looking at the types of computer attacks that occur and then what defenses must be in place.

Module 4, "Internet Security," covers how the Internet works and then identifies the types of risks with using the web, email, and web browsers. It also examines the defenses that can be set up to make the Internet a safer environment when using it for work and play.

Module 5, "Mobile Security," examines some of the attacks on mobile devices and the wireless data networks that support them. First, it explores the types of attacks that a wireless network faces along with the attacks directed at mobile devices using these networks. It also looks at attacks on the mobile devices themselves. It concludes with information on how to protect wireless networks and mobile devices.

Module 6, "Privacy," looks at how businesses are accessing and using our data without our knowledge and permission while earning hundreds of billions of dollars annually. It examines how our data is being taken, who is taking it, and the risks to our privacy. It also details privacy protections that users can take.

Features

To aid you in fully understanding computer and network security, this book includes many features designed to enhance your learning experience.

- **Module Objectives.** Each module begins with a detailed list of the concepts to be mastered within it. This list provides you with both a quick reference to the module's contents and a useful study aid.
- **Cybersecurity Headlines.** Each module opens with the details and explanation of a recent real-world cybersecurity event. This provides a real-world context for applying cybersecurity.
- **Illustrations and Tables.** Numerous illustrations of security vulnerabilities, attacks, and defenses help you visualize security elements and concepts. In addition, the tables provide details and comparisons of practical information.
- **Notes.** Each module's content is supplemented with Note features that provide additional insight and understanding.
- **Cautions.** The Caution features warn you about potential mistakes or problems and explain how to avoid them.
- **Two Rights & a Wrong.** Each section of every module includes a series of questions that can serve as a quick self-assessment of the material. This helps to ensure a comprehensive understanding of the material.
- **Module Summaries.** Each module's text is followed by a summary of the concepts introduced in that module. These summaries provide a helpful way to review the ideas covered in each module.
- **Key Terms.** All of the terms in each module that were introduced with bold text are gathered in a Key Terms list at the end of the module, providing additional review and highlighting key concepts. Key term definitions are included in a Glossary at the end of the text.
- **Review Questions.** The end-of-module assessment begins with a set of review questions that reinforce the ideas introduced in each module. These questions help you evaluate and apply the material you have learned. Answering these questions will ensure that you have mastered the important concepts.
- **Hands-On Projects.** Although it is important to understand the concepts behind security, nothing can improve upon real-world experience. To this end, each module provides several Hands-On Projects aimed at providing you with practical security software and hardware implementation experience. These projects use the Windows 11 operating system, as well as software downloaded from the Internet.
- **Case Projects.** Located at the end of each module are multiple Case Projects. In these exercises, you have the opportunity to "dig deeper" into cybersecurity attacks and defenses and see how they impact our world today.

New to This Edition

- Covers new and updated information on the latest cybersecurity attacks and defenses
- Provides extensive hands-on projects using Microsoft Windows 11
- Incorporates information on protecting Apple macOS computers and Apple iOS devices

- Provides new material on creating a defensive cybersecurity stance
- Includes instructions on using a sandbox environment for performing hands-on projects
- Includes new Cybersecurity Headlines to open each module
- Offers new Hands-On Projects in each module covering some of the latest security software
- Features updated Case Projects in each module

Text and Graphic Conventions

Wherever appropriate, additional information and exercises have been added to this book to help you better understand the topic at hand. Icons throughout the text alert you to additional materials. The icons used in this textbook are described below.

Note 1

The Note icon draws your attention to additional helpful material related to the subject being described.

Caution !

The Caution icon warns you about potential mistakes or problems and explains how to avoid them.

Hands-On Project

Each Hands-On activity in this book is preceded by the Hands-On icon and a description of the exercise that follows.

Case Projects

Case Project icons mark Case Projects, which are scenario-based assignments. In these case examples, you are asked to implement independently what you have learned.

Instructor's Materials

Additional instructor resources for this product are available online. Instructor assets include an Instructor's Manual, PowerPoint® slides, Solutions and Answers Guide, and a test bank powered by Cognero®. Sign up or sign in at **www.cengage.com** to search for and access this product and its online resources.

About the Author

Dr. Mark Ciampa is a Professor of Analytics and Information Systems in the Gordon Ford College of Business at Western Kentucky University in Bowling Green, Kentucky. Previously, he was an Associate Professor and Director of Academic Computing at Volunteer State Community College in Gallatin, Tennessee for 20 years. Mark has worked in the IT industry as a computer consultant for businesses, government agencies, and educational institutions. He has published over 25 articles in peer-reviewed journals and is also the author of over 30 technology textbooks, including

CompTIA CySA+ Guide to Cybersecurity Analyst 2e, CompTIA Security+ Guide to Network Security Fundamentals 7e, CWNA Guide to Wireless LANs 3e, Guide to Wireless Communications, and *Networking BASICS*. Dr. Ciampa holds a Ph.D. in technology management with a specialization in digital communication systems from Indiana State University, and he has certifications in security and healthcare.

Acknowledgments

A large team of dedicated professionals all contributed to this project, and I am honored to be part of such an outstanding group of professionals. First, thanks go to Cengage Product Manager Natalie Onderdonk for providing me the opportunity to work on this project and for her continual support. Thanks also to Cengage Senior Project Managers Jennifer Ziegler and Manoj Kumar for their valuable input. And special recognition again goes to developmental editor Lisa Ruffolo. As always, Lisa took care of all the details so that I could stay focused. And her numerous suggestions and comments were always beneficial and greatly appreciated. It is truly a pleasure to work with Lisa. I also appreciated the significant contributions of the reviewers for this edition: Paul Mercier, Mt. San Antonio College; Amy Osborne, College of Central Florida; Kimberly Perez, Tidewater Community College; and Diego Tibaquirá, EnTec Miami Dade College—Padrón Campus. To everyone on the team, I extend my sincere thanks.

Finally, I want to thank my wonderful wife, Susan. Her interest, support, and love helped me through another project. I could not have done it without her.

Dedication

To Braden, Mia, Abby, Gabe, Cora, Will, and Rowan.

To the User

This book should be read in sequence, from beginning to end. However, each module is a self-contained unit, so after completing Module 1 the reader may elect to move to any subsequent module.

Hardware and Software Requirements

Following are the hardware and software requirements needed to perform the end-of-module Hands-On Projects:

- Microsoft Windows 11 (Projects may also be completed using Windows 10 although the steps may slightly differ.)
- An Internet connection and web browser

Free Downloadable Software Requirements

Free, downloadable software is required for the Hands-On Projects in the following modules:

Module 2

- KeePass

Module 3

- Microsoft Safety Scanner
- ConfigureDefender

Module 4

- Qualys BrowserCheck

Module 5

- NirSoft WifiInfoView

Module 6

- Browzar Private Web Browser, 7-Zip

References

1. "Malware," *AV-Test*, accessed Apr. 8, 2022, https://www.av-test.org/en/statistics/malware/.
2. Boddy, Matt, "Exposed: Cyberattacks on cloud honeypots," *Sophos*, Apr. 9, 2019, accessed June 5, 2019, www.sophos.com/en-us/press-office/press-releases/2019/04/cybercriminals-attack-cloud-server-honeypot-within-52-seconds.aspx.
3. Kress, Robert, "How to develop a cyber-competent boardroom," *Accenture*, Jan. 5, 2022, accessed Apr. 8, 2022, https://www.accenture.com/us-en/blogs/security/cyber-competent-boardroom.

Module 1

Introduction to Cybersecurity

After completing this module, you should be able to do the following:

1. Explain the difficulties in preventing attacks.
2. Define cybersecurity and describe why it is important.
3. Identify threat actors and their attributes.
4. Explain how to build a comprehensive cybersecurity strategy.

Cybersecurity Headlines

Consumers have long benefited from standardized product labeling. Packaged food, automobiles, and appliances are just a few of the examples of items that have these labels. For example, the product label on a bag of chips lists the number of servings per container, calories per serving, total fat, cholesterol, sodium, and any vitamins and minerals. These product labels are useful in knowing what's inside and how it compares with similar products.

Now standardized consumer product labeling is being expanded to cybersecurity.

The National Institute of Standards and Technology (NIST), operating under the U.S. Commerce Department, has been assigned the task to establish criteria that should be the basis for product labels. However, NIST is not designing the label itself, nor is it establishing its own labeling program for consumer software and hardware. Instead, the government initiative is calling for a voluntary approach, and it will be up to the marketplace to determine which organizations might use these cybersecurity labels.

NIST has conducted research and published a set of recommendations on consumer product labeling for cybersecurity. The goals of the NIST approach are to aid consumers in their software purchase decisions by enabling comparisons among products and to educate consumers on security considerations. NIST says that this will also encourage vendors to consider the cybersecurity of their products and help gain consumer trust and confidence.

NIST has tentatively identified three categories of labels:

Binary. Also called a "seal of approval," this is a single label indicating that a product has met a baseline standard. Examples include the Energy Star seal on appliances.

Descriptive. A descriptive (or informational) label provides facts about the properties or features of a product without any grading or evaluation. Examples of descriptive labels include nutrition facts labels found on foods and lighting facts on light bulbs. This information may be displayed in separate ways, such as in tabular format or with icons or text.

Graded. A graded (tiered) label indicates the degree to which a product has satisfied a specific standard. It can be based on attaining increasing levels of performance against specified criteria. These grades may be represented by colors (red-yellow-green), numbers of icons (stars or security shields), or other metaphors (gold-silver-bronze). One example is vehicle safety ratings.

A layered label approach involves one of the three types of labels initially presented to the consumer with additional information offered online. This may include a web link reference to a website or a Quick Response (QR) code

Continued

that takes a consumer to more detailed information. NIST is proposing a layered binary label. They are also proposing a "robust consumer education campaign" to educate users on the labels.

Soon consumers may be able to use product labels to help them determine the cybersecurity strength of hardware devices and software before making a purchase.

Did you hear that the maker of the game "Axie Infinity" reported that attackers had stolen from it over $500 million worth of cryptocurrency, one of the biggest thefts in the history of digital currency?[1] Or that patients sued Logan Health in the wake of a breach that exposed the personal and health data of nearly 214,000 patients, and this breach was the second against the health system in less than three years?[2] Or that the British toy retailer The Works shut down five stores after attackers gained access to its computer systems that caused problems with its checkout systems?[3] Or that Mailchimp, which sells newsletter and marketing software, was the victim of an attack that stole information on hundreds of its finance and cryptocurrency customers, forcing a customer, a crypto wallet firm, to entirely shut down one of its web domains?[4]

And all of this happened within *the same week*?

You may not have heard of any of these incidents. While in the past just one of these cyber events would have been newsworthy headlines that immediately went viral across the Internet, today they barely register a blip on the radar screen. It's not because they are unimportant: rather, it's simply because cybersecurity attacks have become so commonplace that you hardly notice them any longer. *Oh, there was another data breach today? So, what else is new?*

The sheer volume of attacks has reached astronomical proportions. The AV-TEST Institute receives instances of over 450,000 new malicious programs (malware) and potentially unwanted applications (PUA) *each day*. The total number of instances of malware has grown from 182 million in 2013 to over 1.34 billion today.[5] Cybercrime has been called the "greatest transfer of economic wealth in history," and it is estimated that it could reach $10.5 *trillion annually* by 2025.[6] And the dismal numbers go on and on.

But it is not just large multinational corporations that are the targets of attackers. Anyone who uses a computing device—like a laptop, tablet, or smartphone—or who has a device that is connected to the Internet—like a security camera, doorbell, or thermostat—is at risk of being attacked and compromised.

However, although attacks are at an all-time high, the overwhelming majority of users still struggle with how to actually make their devices secure. Ask yourself this question: If you were warned that a vicious new attack was to begin within the next 15 minutes, would you know the steps to take to check the security on your devices, and quickly fix anything deficient? Only a small handful of users could truthfully answer *Yes* to that question.

It is important for all computer users today to be knowledgeable about computer cybersecurity and to be able to take the necessary steps to defend against attacks. Applying practical security in your world has never been more important than it is right now.

This module introduces you to cybersecurity. It begins by examining why it is so difficult to protect our devices. It then describes cybersecurity in more detail and explores why it is important. The module also looks at who is behind these attacks and then concludes by looking at the steps necessary for building a comprehensive cybersecurity strategy.

Difficulties in Preventing Attacks

Why is it so hard to prevent attacks? Can't we have a single software program to thwart all attacks, or a hardware device to block any intruders, or a single button to click that spreads a protective dome over all the devices in a house or apartment?

Unfortunately, the answer is *No*. None of these exist—and they never will.

Far from being simple, preventing attacks is hard—*very hard*. That is because there are many difficulties associated with countering attacks, which include the following:

- **Universally connected devices**. It is unthinkable today for any technology device—not only a computer or tablet but also a smartphone, door lock, or doorbell camera—not to be connected to the Internet. Although this connectivity provides enormous benefits, it also makes it easy for an attacker halfway around the world to silently launch an attack against any connected device.

- **Increased speed of attacks**. Attackers can quickly scan millions and millions of devices to find weaknesses and then immediately launch attacks with unprecedented speed. And these attack tools will "scan and attack" completely on their own without any human help, thus increasing the speed at which users are attacked.
- **Greater sophistication of attacks**. Attacks are becoming more complex, making it difficult to detect and defend against them. Attackers today use common Internet communications and web applications to hide their attacks, making it more difficult to distinguish an attack from legitimate network traffic. Other attack tools even vary their behavior, so the same attack appears differently each time, further complicating detection.
- **Availability and simplicity of attack tools**. Whereas at one time an attacker needed extensive technical knowledge of networks and computers as well as the ability to write a software program to generate an attack, that is no longer the case. Today's software attack tools require little if any sophisticated knowledge on the part of the attacker. In fact, many of the tools, such as the Kali Linux menu shown in Figure 1-1, allow the user to simply select options from a menu. And these tools are freely available for anyone to download and use.

Figure 1-1 Menu of attack tools

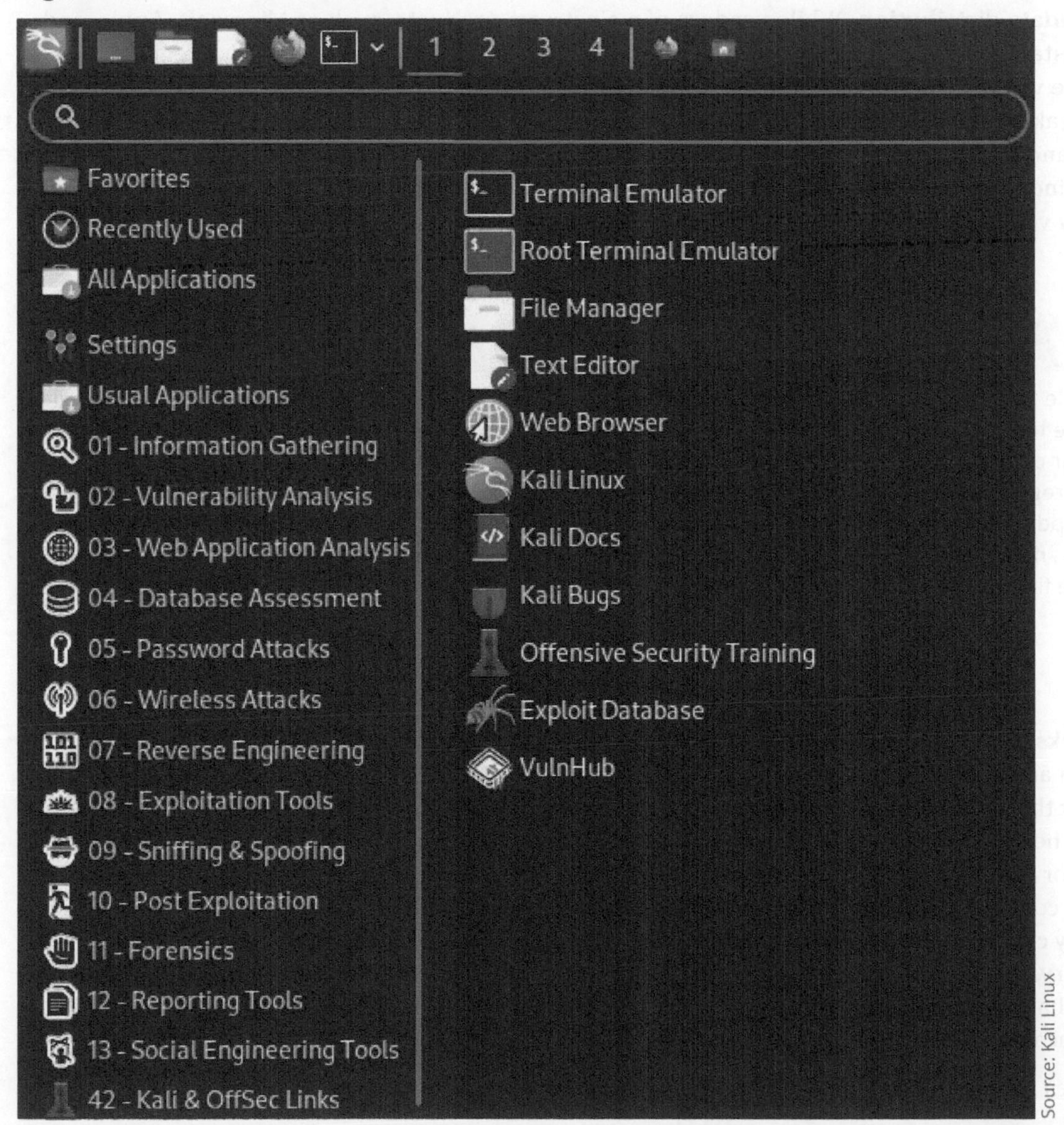

Source: Kali Linux

Note 1

Usually these tools contain disclaimers such as *"These tools are provided so that you can test your system against threats, understand the nature of these threats, and protect your own systems from similar attacks. At no time should they be used against another system that you do not own or control. Using these tools in an unauthorized fashion may be illegal in your jurisdiction and may be considered as a terms of service violation or even professional misconduct."* However, there is no means to prevent anyone from using the attack tool however they wish.

- **Faster detection of weaknesses**. The exposure of a security weakness in hardware and software can be quickly uncovered and then exploited. Often attackers discover these weaknesses even before anyone else knows that they even exist.
- **Delays in security updating**. Hardware and software vendors who develop and sell products are simply overwhelmed trying to keep pace with updating their products against attacks. As noted earlier, one antivirus software security institute receives more than 450,000 submissions of potential malware *each day*. At this rate, some vendors would have to create and distribute updates to all users *every few seconds* to keep them fully protected. Any delay in distributing security updates adds to the difficulties in preventing attacks.
- **Weak security update distribution**. While vendors of mainstream products, such as Microsoft, Apple, and Adobe, have an established system for distributing security updates for their products on a regular basis, few other software vendors have invested in these costly distribution systems. Users are generally unaware that a security weakness has been found or an update even exists because vendors have no reliable means to alert the user and provide the update. Some vendors do not even create security updates that fix problems, but instead incorporate the update into an entirely new version of the product—and then require users to pay for the new version.

Note 2

While all users of Apple smartphones regularly receive security updates, those who own phones running Google Android may not. Apple has developed and sells both the hardware and software for their phones and has complete control over the regular distribution of updates. In contrast, Android runs on smartphones from many different vendors. Although Google regularly makes security updates available, it is up to the phone's vendor and even the wireless carrier when to send updates to users. This results in a very uneven update distribution: some newer Android smartphones receive updates monthly while others get them only each quarter. And some Android vendors only distribute security updates for the first two years of the device's life, while Apple usually supports their smartphones with updates for up to eight years.

- **Distributed attacks**. Attackers can use hundreds of thousands or even millions of computers under their remote control in an attack against a single device. This overwhelming onslaught makes it impossible to identify and block the source to stop the attack.
- **User confusion**. The one factor that undoubtedly accounts for the greatest difficulty in preventing attacks is user confusion. For many years, end-users have been called upon to make often difficult security decisions and then perform complicated procedures on their devices—often with little information to guide them. This is compounded by cybersecurity information circulated through consumer news outlets and websites that is often contradictory, inaccurate, or misleading, resulting in even more user confusion. Although it is universally recognized that instruction and training on cybersecurity are critical, few schools offer comprehensive cybersecurity education to all students appropriate to their age.

Table 1-1 summarizes the reasons why it is difficult to defend against today's attacks.

Table 1-1 Difficulties in defending against attacks

Reason	Description
Universally connected devices	Attackers from anywhere in the world can send attacks.
Increased speed of attacks	Attackers can launch attacks against millions of computers within minutes.
Greater sophistication of attacks	Attack tools vary their behavior so the same attack appears differently each time.
Availability and simplicity of attack tools	Attacks are no longer limited to highly skilled attackers.
Faster detection of weaknesses	Attackers can discover security holes in hardware or software more quickly.
Delays in security updating	Vendors are overwhelmed trying to keep pace by updating their products against the latest attacks.
Weak security update distribution	Many software products lack the means to distribute security updates in a timely fashion.
Distributed attacks	Attackers use thousands of computers in an attack against a single computer or network.
User confusion	Users are required to make difficult security decisions with little or no instruction.

Two Rights & A Wrong

1. Some attacks can vary their behavior so that the same attack appears differently.
2. Attack tools still require a high degree of skill and knowledge to use them.
3. The single difficulty that accounts for the greatest difficulty in preventing attacks is the increased speed of attacks.

See Appendix A for the answer.

What Is Cybersecurity?

A first step in the study of cybersecurity is to define exactly what it is. This involves knowing the definition of security and how it relates to cybersecurity. It also includes becoming familiar with the terminology used in this area. This helps to understand the importance of cybersecurity.

Understanding Security

What is *security*? The word comes from the Latin, meaning *free from care*. Sometimes security is defined as *the state of being free from danger*, which is the *goal* of security. It is also defined as the *measures taken to ensure safety*, which is the *process* of security. Since complete security can never be fully achieved, the focus of security is more often on the process instead of the goal. In this light security can be defined as *the necessary steps to protect from harm*.

It is important to understand the relationship between *security* and *convenience*. The relationship between these two is not *directly proportional* (as security is increased, convenience is increased). Instead, it is completely the opposite, known as *inversely proportional* (as security is increased, convenience is decreased). In other words, the more secure something becomes, the less convenient it may become to use. These are illustrated in Figure 1-2.

Note 3

In addition, as convenience is increased usually security is decreased.

Consider an app on a smartphone in which the user must enter a password but then also wait for a special code to be delivered and then entered. The additional step of entering the special code may be less convenient than just entering the password by itself, but it affords greater security. Thus, as security is increased usually convenience is decreased. Security is often described as *sacrificing convenience for the sake of safety*.

Figure 1-2 Relationship of security to convenience

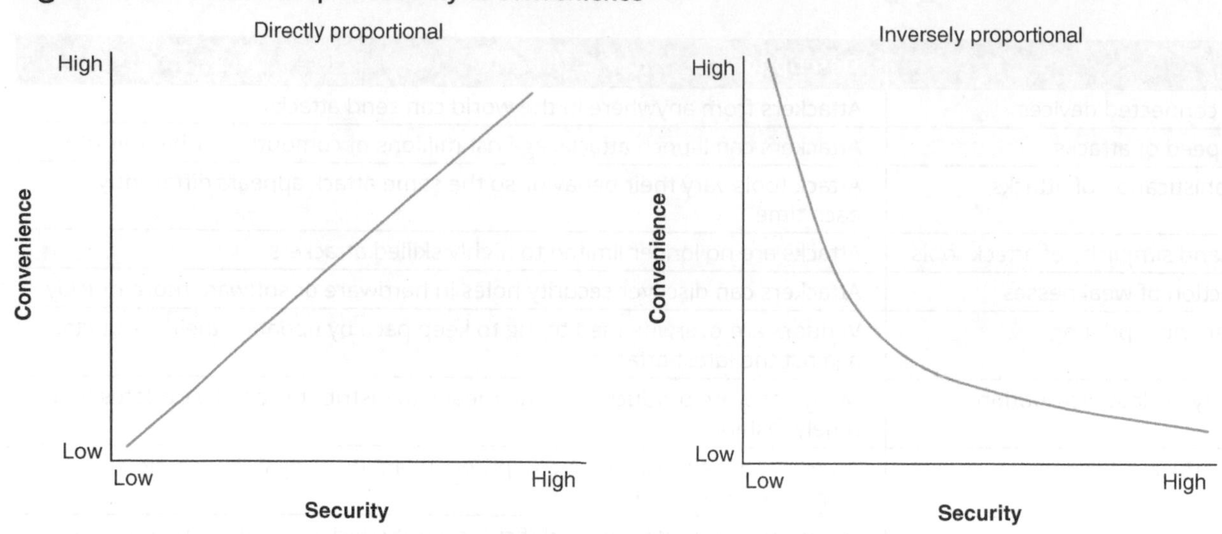

Defining Cybersecurity

Several terms are used when describing security for technology: *computer security, information security, IT security,* and *information assurance*, to name just a few. Whereas each has its share of proponents and slight variations of meanings, the term *cybersecurity* may be the most relevant.

Cybersecurity has been called both an *art* and a *practice*. An *art* is the systematic application of knowledge and skills, while a *practice* is a repeated or continual action.

Cybersecurity as an Art

Cybersecurity is sometimes said to be the *art*—systematically applying the correct knowledge and skills—of protecting networks and devices. As such, it illustrates that cybersecurity is never a "one and done" event but requires continual effort. This effort requires a degree of knowledge of what to do and the skills to do it.

However, even the best cybersecurity knowledge and skills cannot completely prevent all attacks or guarantee that a system will be totally secure, just as the security measures taken for a house can never guarantee complete safety from a burglar. The goal of cybersecurity is to ensure that protective measures are properly implemented to ward off attacks, prevent the total collapse of the system when a successful attack does occur, and recover as quickly as possible. Thus, the first element of cybersecurity is *protection*.

Cybersecurity as a Practice

Cybersecurity is also a *practice*, or a repeated and continual action. This action is to ensure the confidentiality, integrity, and availability of *information* (the second element) that provides value to people and enterprises. Three protections must be extended over information:

1. **Confidentiality**. It is important that only approved individuals can access sensitive information. For example, the credit card number used to make an online purchase must be kept secure and not made available to other parties. **Confidentiality** ensures that only authorized parties can view the information. Providing confidentiality can involve several different security tools, ranging from software to encrypt the credit card number stored on the web server to door locks to prevent access to those servers.
2. **Integrity**. **Integrity** ensures that the information is correct and no unauthorized person or malicious software has altered the data. In the example of an online purchase, an attacker who could change the amount of a purchase from $10,000.00 to $1.00 would violate the integrity of the information.
3. **Availability**. Information has value if the authorized parties who are assured of its integrity can access the information. **Availability** ensures that data is accessible to only authorized users and not to unapproved individuals. In this example, the total number of items ordered as the result of an online purchase must be made available to an employee in a warehouse so that the correct items can be shipped to the customer but not available to a competitor.

In addition to these three protections, another set of protections must be implemented to secure information:

1. **Authentication**. **Authentication** ensures that the individual is who she claims to be (the authentic or genuine person) and not an imposter. A person accessing the web server that contains a user's credit card number must prove that she is indeed who she claims to be and not a fraudulent attacker. One way in which authentication can be performed is by the person providing a password that only she knows.
2. **Authorization**. **Authorization** is providing permission or approval to specific technology resources. After a person has provided authentication, she may have the authority to access the credit card number or enter a room that contains the web server, provided she has been given prior authorization.
3. **Accounting**. **Accounting** provides tracking ("audit trail") of events. This may include a record of who accessed the web server, from what location, and at what specific time.

Thus, cybersecurity describes the task of securing information that is in a digital form. Yet how is the information protected? The digital information may be manipulated by a microprocessor (such as on a computer), preserved on a storage device (like a hard drive or USB flash drive), or transmitted over a network (such as the Internet). Because this information is stored on computer hardware, manipulated by software, and transmitted by communications, each of these areas must be sheltered. The third element of cybersecurity is to protect the integrity, confidentiality, and availability of information *on the devices that store, manipulate, and transmit the information.*

You must remember that the networks and devices are not the ultimate goals of protection. Rather, it is the information (in digital form) on these networks and devices that must be secured from unauthorized access or criminal use. That is because the information on the device is many times more valuable than the device itself.

Note 4

Consider for a moment your smartphone. If you were to lose your smartphone, you could likely replace it with another phone for several hundred dollars. But the information on your smartphone—the pictures of your most recent vacation, your list of contacts, those special text messages—are "priceless" to most users. This illustrates that the information stored on a device is of higher worth than the device itself. Protecting information is the goal of cybersecurity, and this is achieved through protecting the devices on which the information resides.

Information security is achieved through a process that combines three entities. As shown in Figure 1-3 and Table 1-2, information and the hardware, software, and communications are protected in three layers: products, people, and policies and procedures. These three layers interact with each other: procedures enable people to understand how to use products to protect information.

Table 1-2 Information security layers

Layer	Description
Products	Forms the security around the data. May be as basic as door locks or as complicated as network security equipment.
People	Those who implement and properly use security products to protect data.
Policies and procedures	Plans and policies established by an organization to ensure that people correctly use the products.

A comprehensive definition of cybersecurity involves both the goals and process, as well as the art and the practice. **Cybersecurity** may be defined as *that which protects the integrity, confidentiality, and availability of information on the devices that store, manipulate, and transmit information through products, people, and procedures.*

Caution 

Cybersecurity should not be viewed as a war to be won or lost. Just as crimes such as burglary can never be completely eradicated, neither can attacks against technology devices. The goal is not a complete victory but maintaining equilibrium: as attackers take advantage of a weakness in a defense, defenders must respond with an improved defense. Cybersecurity is an endless cycle between attacker and defender.

Figure 1-3 Information security layers

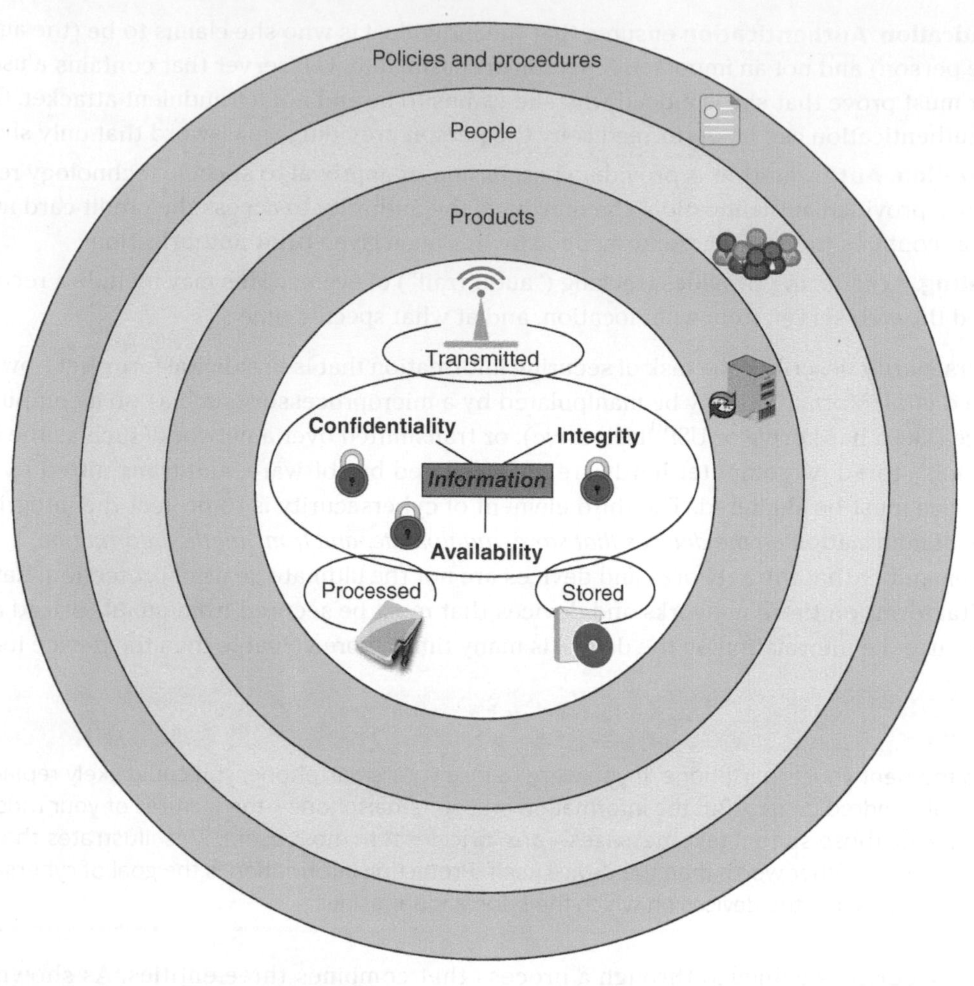

Cybersecurity Terminology

As with many advanced subjects, cybersecurity has its own set of terminology. The following scenario helps to illustrate cybersecurity terms and how they are used.

Suppose that Cora wants to purchase a new motorized Italian scooter to ride from her apartment to school and work. However, because several scooters have been stolen near her apartment, she is concerned about its protection. Although she parks the scooter in the gated parking lot in front of her apartment, a hole in the fence surrounding the apartment complex makes it possible for someone to access the parking lot. Cora's scooter and the threat to it are illustrated in Figure 1-4.

Cora's new scooter is an **asset**, which is defined as an item that has value. Cora is trying to protect her scooter from a **threat**, which is a type of action that has the potential to cause harm. Cybersecurity threats are events or actions that represent a danger to information assets. A threat by itself does not mean that security has been compromised; rather, it simply means that the potential for creating a loss is real.

Figure 1-4 Cybersecurity terminology analogy

Stolen scooter (risk)
Loss of scooter (threat)
Exploit (go through fence hole)
Fence hole (vulnerability)
Thief (threat agent)
Scooter (asset)
annet999/Shutterstock.com

A **threat agent** is a person or element that has the power to carry out a threat. For Cora, the threat agent is a thief. In cybersecurity, a threat agent could be a person attempting to break into a secure computer network.

Cora wants to protect her scooter and is concerned about a hole in the fencing around her apartment. The hole in the fencing is a **vulnerability**, which is a weakness that can be exploited. An example of a vulnerability that cybersecurity must deal with is a software weakness in an operating system that allows an unauthorized user to gain control of a computer without the user's knowledge or permission.

Note 5

The word *vulnerability* comes from Latin meaning *wound*.

If a thief can get to Cora's scooter because of the hole in the fence, then that thief is taking advantage of the vulnerability. This is known as exploiting the vulnerability through an **attack vector**, which is a pathway used by a threat actor to penetrate a system. An attacker, knowing that a flaw in a web server's operating system has not been patched, can use this attack vector to exploit the vulnerability to steal user passwords.

Note 6

Email is a popular attack vector: almost 94 percent of all malware is delivered through email to an unsuspecting user.[7]

Cora must make a decision: what is the probability (**threat likelihood**) that the threat will come to fruition and her scooter stolen? This can be understood in terms of risk. A **risk** is a situation that involves exposure to some type of danger. Cora could take different options regarding the risk of her scooter being stolen. Cora could decide based on the risk of the scooter being stolen that she will not purchase the new scooter (*risk avoidance*). Or she could accept the risk and buy the new scooter, knowing there is the chance of it being stolen by a thief entering through a hole in the fence (*risk acceptance*). She might complain to the apartment manager about the hole in the fence in order to have it repaired and make the risk less serious (*risk mitigation*) or ask the manager to post signs that said, "Trespassers will be punished to the full extent of the law" (*risk deterrence*). What Cora is most likely to do is purchase insurance so that the insurance company absorbs the loss and pays her if the scooter is stolen, in essence making someone else responsible (*risk transference*).

Table 1-3 summarizes these cybersecurity terms.

Table 1-3 Cybersecurity terminology

Term	Example in Cora's scenario	Example in information security
Asset	Scooter	Employee database
Threat	Steal scooter	Steal data
Threat agent	Thief	Attacker, hurricane
Vulnerability	Hole in fence	Software defect
Attack vector	Climb through hole in fence	Access web server passwords through software flaw
Threat likelihood	Probability of scooter stolen	Likelihood of virus infection
Risk	Not purchase scooter	Not install wireless network

Why Cybersecurity Is Important

Cybersecurity is important to individuals as well as organizations. That is because it can help prevent data theft, thwart identity theft, foil cyberterrorism, avoid the legal consequences of not securing information, and maintain productivity.

Prevent Data Theft

Security is most often associated with *theft prevention*: Cora could park her scooter in a locked garage in order to prevent it from being stolen. The same is true with cybersecurity: preventing data from being stolen is often cited as a primary objective. For a business, it is necessary to guard against data theft because attackers are eager to steal proprietary business information, such as research for a new product or a list of customers.

Individuals as well are often the targets of data thievery. One type of personal data that is a prime target of attackers is payment card numbers, such as debit cards, credit cards, or gift cards. These stolen numbers are often sold on an online marketplace to other criminals. They can be used to purchase thousands of dollars of merchandise online—without having the actual card—before the victim or bank is even aware the number has been stolen. Some of the common techniques used by payment card thieves include:

- Thieves typically determine if a stolen card number is still active by making a small purchase, which is unlikely to generate the attention of the user or the bank that issued the card.
- Some online marketplace sellers will provide a guarantee that the stolen card numbers will remain active for a specific period of time or for the purchase of a minimum amount of merchandise before the card number is revoked.
- Sellers will often monitor how their customers use the stolen cards in order to ensure they do not generate too much attention and thus risk being discovered, which would then prevent other customers who have purchased similar cards from being able to make purchases.
- Stolen card numbers that also include personal information such as the birth date and Social Security number of the cardholder are worth more than just the card number itself. Thieves can use this information to uncover other personal information about the victim and thus be in a better position to answer security challenge questions that might be asked by the bank if a large purchase is being made.

Thwart Identity Theft

Identity theft involves stealing another person's personal information, such as a Social Security number, and then using the information to impersonate the victim, often for financial gain. The thieves may create new bank or credit card accounts under the victim's name and then charge large purchases to these accounts, leaving the victim responsible for the debts and ruining her credit rating.

Note 7

In some instances, thieves have bought cars and even houses by taking out loans in someone else's name.

The following statistics help illustrate the breadth of identity theft:

- Each year losses from reported identity theft exceed $3 billion.
- In one year, 47 percent of Americans experienced financial identity theft, and 30 percent have experienced identity theft more than once.
- While seniors over 60 years old are the most common victims of identity theft, Millennials account for 44 percent of U.S. identity fraud reports.
- A new victim of identity theft occurs every 2 seconds.[8]

One rapidly growing area of identity theft involves identity thieves filing fictitious income tax returns with the U.S. Internal Revenue Service (IRS) to receive the victim's tax refund. Each year several billion dollars in refund checks are sent to identity thieves who filed fraudulent tax returns and denied victims their returns.

Note 8

Like stolen credit cards, stolen personal information is usually posted on an online marketplace for purchase and use by other criminals. One major online marketplace posted hundreds of databases of stolen data containing more than 10 billion unique records for individuals from around the world. The information included sensitive personal and financial information, such as stolen bank routing and account numbers, credit card information, login credentials, and Social Security numbers. Specialized databases were also offered, with descriptions such as "Usernames and passwords for online customer accounts issued by a major broadcasting and cable company in the U.S." and private account information from "A major telecommunications company and wireless network operator that provides services in the U.S." This site was eventually taken down by law enforcement.

Foil Cyberterrorism

The FBI defines **cyberterrorism** as any "premeditated, politically motivated attack against information, computer systems, computer programs, and data which results in violence against noncombatant targets by subnational groups or clandestine agents."[9] Unlike an attack that is designed to steal, cyberterrorism attacks are intended to cause panic or provoke violence among citizens. Attacks are directed at targets such as the banking industry, military installations, power plants, air traffic control centers, and water systems. These are desirable targets because they can significantly disrupt the normal activities of a large population. For example, disabling an electrical power plant could cripple businesses, homes, transportation services, and communications over a wide area.

Note 9

One of the challenges in combating cyberterrorism is that many of the prime targets are not owned and managed by the federal government. Because these are not centrally controlled, it is difficult to mandate, coordinate, and maintain security.

The threat of cyberterrorism is considered serious and increasing. In one year, 118 "significant cyberattacks" occurred that were linked to cyberterrorism.[10]

Avoiding Legal Consequences

Several international, national, and state laws have been enacted to protect the privacy of electronic data. Businesses that fail to protect data they possess from cybercriminals may face both legal actions as well as significant financial penalties.

Those companies that perform business in the European Union (EU) are legally obligated under the **General Data Protection Regulation (GDPR)** to inform the EU's Information Commissioner's Office (ICO) if they suffer a breach involving the personal information of customers or employees.

Note 10

The GDPR not only applies to organizations that are located within the EU but also applies to organizations located outside of the EU if they offer goods or services to EU citizens or monitor the behavior of EU data that pertains to its subjects. The GDPR also applies to all companies processing and holding the personal data of subjects residing in the EU regardless of the company's location.

In the United States, some of the federal laws include the following:

- **The Health Insurance Portability and Accountability Act of 1996 (HIPAA).** Under the **Health Insurance Portability and Accountability Act (HIPAA)**, healthcare enterprises must guard protected healthcare information and implement policies and procedures to safeguard it, whether it be in paper or electronic format. Those who wrongfully disclose individually identifiable health information can be fined up to $50,000 for each violation up to a maximum of $1.5 million per calendar year and sentenced to up to 10 years in prison.

Note 11

HIPAA regulations have been expanded to include all third-party "business associate" organizations that handle protected healthcare information. Business associates are defined as any subcontractor that creates, receives, maintains, or transmits protected health information on behalf of a covered HIPAA entity. These associates must now comply with the same HIPAA security and privacy procedures.

- **The Sarbanes-Oxley Act of 2002 (Sarbox).** As a reaction to a rash of corporate fraud, the **Sarbanes-Oxley Act (Sarbox)** is an attempt to fight corporate corruption. Sarbox covers the corporate officers, auditors, and attorneys of publicly traded companies. Stringent reporting requirements and internal controls on electronic

financial reporting systems are required. Corporate officers who willfully and knowingly certify a false financial report can be fined up to $5 million and serve 20 years in prison.
- **The Gramm-Leach-Bliley Act (GLBA).** Like HIPAA, the **Gramm-Leach-Bliley Act (GLBA)** passed in 1999 protects private data. GLBA requires banks and financial institutions to alert customers of their policies and practices in disclosing customer information. All electronic and paper data containing personally identifiable financial information must be protected. The penalty for noncompliance for a class of individuals is up to $500,000.
- **Payment Card Industry Data Security Standard (PCI DSS).** The **Payment Card Industry Data Security Standard (PCI DSS)** is a set of security standards that all companies that process, store, or transmit credit card information must follow. PCI applies to any organization or merchant, regardless of its size or number of card transactions, that processes transactions either online or in person. The maximum penalty for not complying is $100,000 per month.

However, the United States does not have a comprehensive federal law that requires notification in the event of a data breach. In that absence, all states have their own laws requiring that victims be notified if their personal information has been stolen. However, no two state laws are the same, often resulting in confusion. Table 1-4 lists some of the differences between state laws.

Table 1-4 Differences in state laws

Description	Example state	Explanation
Broader definition of personal information	Alabama	A tax identification number; passport number; military identification number; other unique identification number issued on a government document used to verify the identity of a specific individual; any information about an individual's medical history, mental or physical condition, or medical treatment or diagnosis by a healthcare professional; health insurance policy number or subscriber identification number and any unique identifier used by a health insurer to identify the individual; a username or email address in combination with a password or security question and answer.
Notification triggered by access to data and not documented theft	Florida	"Breach of security" means unauthorized access to personal information in electronic format.
Breach must satisfy risk-of-harm analysis	Arkansas	Notification is not required if, after a reasonable investigation, the business determines that there is no reasonable likelihood of harm to customers.
Expanded notification beyond impacted citizens	Colorado	Additional notice must be provided to the state attorney general.
Includes encryption safe harbor	Alaska	The statute only applies to unencrypted information or encrypted information when the encryption key has also been disclosed.
Covers other forms of data	Hawaii	The statute applies to personal information in any form, whether computerized, paper, or otherwise.

Maintaining Productivity

Cleaning up after an attack by a business diverts time, money, and other resources away from normal activities. Users who are victims of a successful attack may spend days restoring their computers to the state before the attack, or they may have to pay a technology professional to complete the task. During this time, the computer is unavailable and personal productivity suffers.

Employees of an organization likewise are impacted due to an attack that renders their device useless. These workers cannot be productive and complete important tasks during or after an attack because computers and networks cannot function properly. Table 1-5 provides a sample estimate of the lost wages and productivity during an attack and the subsequent cleanup.

Table 1-5 Cost of attacks

Number of total employees	Average hourly salary	Number of employees to combat attack	Hours required to stop attack and clean up	Total lost salaries	Total lost hours of productivity
100	$25	1	48	$4,066	81
250	$25	3	72	$17,050	300
500	$30	5	80	$28,333	483
1,000	$30	10	96	$220,000	1293

Two Rights & A Wrong

1. Cybersecurity is directly proportional to convenience.
2. Confidentiality ensures that only authorized parties can view the information.
3. Authentication ensures that the individual is who she claims to be and not an imposter.

See Appendix A for the answer.

Who Are the Attackers?

In the past, the term **hacker** referred to a person who used advanced computer skills to attack computers. Yet because that title often carried with it a negative connotation, it was qualified in an attempt to distinguish between different types of these attackers. These types are summarized in Table 1-6.

Table 1-6 Types of hackers

Hacker type	Description
Black hat hackers	Attackers who violate computer security for personal gain (such as stealing credit card numbers) or to inflict malicious damage (corrupting a hard drive).
White hat hackers	Also known as *ethical attackers*, they attempt to probe a system (with an organization's permission) for weaknesses and then privately provide that information back to the organization.
Gray hat hackers	Attackers who attempt to break into a computer system without the organization's permission (an illegal activity) but not for their own advantage; instead, they publicly disclose the attack in order to shame the organization into taking action.

However, as cybersecurity attacks have changed, so too have the terms used to refer to attackers. Today a **threat actor** is a term used to describe individuals or entities who are responsible for cyber incidents against the technology equipment of enterprises and users. The generic term *attacker* is also used.

Threat actors are classified in more distinct categories, such as cybercriminals, script kiddies, brokers, insiders, cyberterrorists, hactivists, and state actors.

Cybercriminals

The very first cyberattacks that occurred were mainly for the attackers to show off their technology skills (*fame*). However, that soon gave way to attackers with the focused goal of financial gain (*fortune*). Often those that attack for financial gain are called **cybercriminals**.

Financial cybercrime is often divided into three categories based on different targets:

- **Individual users**. The first category focuses on individuals as the victims. The threat actors steal and use stolen data, credit card numbers, online financial account information, or Social Security numbers to profit from its victims or send millions of spam emails to peddle counterfeit drugs, pirated software, fake watches, and pornography.

- **Enterprises**. The second category focuses on enterprises and business organizations. Threat actors attempt to steal research on a new product from an enterprise so that they can sell it to an unscrupulous foreign supplier who will then build an imitation model of the product to sell worldwide. This deprives the legitimate business of profits after investing hundreds of millions of dollars in product development, and because these foreign suppliers are in a different country, they are beyond the reach of domestic enforcement agencies and courts.
- **Governments**. Governments are also the targets of threat actors. If the latest information on a new missile defense system can be stolen, it can be sold—at a high price—to that government's enemies. In addition, government information is often stolen and published to embarrass the government before its citizens and force it to stop what is considered a nefarious action.

Script Kiddies

Script kiddies are typically younger individuals who want to attack computers, yet they lack the knowledge of computers and networks needed to do so. Script kiddies instead do their work by downloading automated attack software (scripts) from websites and using it to perform malicious acts. Figure 1-5 illustrates the skills needed for creating attacks. Over 40 percent of attacks require low or no skills and are frequently conducted by script kiddies.

Figure 1-5 Skills needed for creating attacks

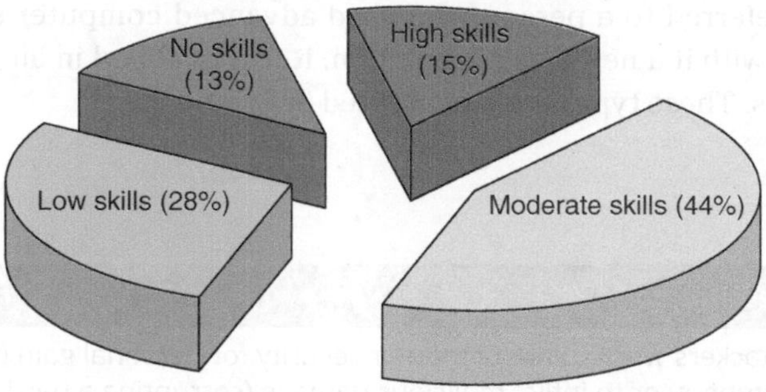

Brokers

In recent years, several software vendors have started financially rewarding individuals who uncover vulnerabilities in their software and then privately report it back to the vendors so that the weaknesses can be addressed. Some vendors even sponsor annual competitive contests or "Bug Bounties" that handsomely pay those who can successfully attack their software to reveal vulnerabilities. The vendors then use this knowledge to remove the vulnerability.

Note 12

The payouts for bug bounties vary widely, with most payouts averaging between $200 and $2,000 per bug disclosure. Google advertises that will pay up to $31,337 for a specific type of reported bug.[11] However, Apple says that it will pay up to $1 million for a "Zero-click kernel code execution with persistence and kernel PAC bypass" bug.[12]

However, other individuals who uncover vulnerabilities do not report them to the software vendor but instead sell them to the highest bidder. Known as **brokers**, these attackers sell their knowledge of a vulnerability to other attackers or even governments. These buyers are generally willing to pay a high price because this vulnerability is unknown to the software vendor and thus is unlikely to be "patched" until after new attacks based on it are already widespread.

Insiders

Another serious threat to an organization actually comes from an unlikely source: its employees, contractors, and business partners, often called **insiders**. For example, a healthcare worker disgruntled over an upcoming job termination might illegally gather health records on celebrities and sell them to the media, or a securities trader who loses billions of dollars on bad stock bets could use her knowledge of the bank's computer security system to conceal the losses through fake transactions. It is reported that 60 percent of breaches were attributed to insiders who abused their right to access corporate information.[13] These attacks are harder to recognize because they come from within the organization yet may be more costly than attacks from the outside.

Most malicious insider attacks consist of the sabotage or theft of intellectual property. Sabotage often comes from employees who have announced their resignation or have been formally reprimanded, demoted, or fired. When theft is involved, the offenders are usually salespeople, engineers, computer programmers, or scientists who actually believe that the accumulated data is owned by them and not the organization (most of these thefts occur within 30 days of the employee resigning). In some instances, the employees are moving to a new job and want to take "their work" with them, while in other cases the employees have been bribed or pressured into stealing the data.

Note 13

Employees have even been pressured into stealing from their employer through blackmail or the threat of violence.

Cyberterrorists

Terrorism today has expanded from planting bombs or other acts of violence against innocent civilians to cyberattacks on a nation's network and computer infrastructure. Yet the goals are the same: to cause disruption and panic among citizens. Known as **cyberterrorists**, their motivation is ideological, attacking for the sake of their principles or beliefs. Cyberterrorists are highly feared, for it is almost impossible to predict when or where an attack may occur. Unlike cybercriminals who continuously probe systems or create attacks, cyberterrorists can be inactive for several years and then suddenly strike in a new way. Their targets may include a small group of computers or networks that can affect the largest number of users, such as the computers that control the electrical power grid of a state or region.

Hactivists

Another group that is strongly motivated by ideology (for the sake of their principles or beliefs) is **hactivists** (a combination of the words *hack* and *activism*). Most hactivists do not explicitly call themselves "hactivists" but the term is commonly used by security researchers and journalists to distinguish them from other types of threat actors.

In the past, the types of attacks by hactivists often involved breaking into a website and changing its contents as a means of making a political statement. (One hactivist group changed the website of the U.S. Department of Justice to read *Department of Injustice*.) Other attacks were retaliatory: hactivists have disabled the website belonging to a bank because that bank stopped accepting online payments that were deposited into accounts belonging to groups that were supported by hactivists. Today, many hactivists work through disinformation campaigns by spreading fake news and supporting conspiracy theories.

Note 14

Hactivists were particularly active during the coronavirus disease (COVID-19) pandemic. One large group of what were considered far-right neo-Nazi hactivists embarked on a months-long disinformation campaign designed to weaponize the pandemic by questioning scientific evidence and research. In another instance, thousands of breached email addresses and passwords from U.S. and global health organizations, including U.S. National Institutes of Health, Centers for Disease Control and Prevention, and the World Health Organization were distributed on Twitter by these groups to harass and distract these health organizations.

State Actors

Instead of using an army to march across the battlefield to strike an adversary, governments are increasingly employing their own state-sponsored attackers for launching cyberattacks against their foes. These are known as **state actors**. Their foes may be foreign governments or even citizens of their own nation that the government considers hostile or threatening. A growing number of attacks from state actors are directed toward businesses in foreign countries with the goal of causing financial harm or damage to the enterprise's reputation.

Many security professionals believe that state actors might be the deadliest of any threat actors. When fortune motivates a threat actor, but the target's defenses are too strong, the attacker simply moves on to another promising target with less-effective defenses. With state actors, however, the target is very specific, and the attackers keep working until they are successful. This is because state-sponsored attackers are highly skilled and have enough government resources to breach almost any security defense.

State actors are often involved in multiyear intrusion campaigns targeting highly sensitive economic, proprietary, or national security information. This has created a new class of attacks called **Advanced Persistent Threat (APT)**. These attacks use innovative attack tools (*advanced*) and once a system is infected, they silently extract data over an extended period of time (*persistent*). APTs are most commonly associated with state actors.

Table 1-7 lists several characteristics of these different attackers.

Table 1-7 Characteristics of attackers

Attacker category	Objective	Typical target	Sample attack
Cybercriminals	Fortune over fame	Users, businesses, governments	Steal credit card information
Script kiddies	Thrills, notoriety	Businesses, users	Erase data
Brokers	Sell vulnerability to highest bidder	Any	Find vulnerability in operating system
Insiders	Retaliate against employer, shame government	Governments, businesses	Steal documents to publish sensitive information
Cyberterrorists	Cause disruption and panic	Businesses	Cripple computers that control water treatment
Hactivists	Right a perceived wrong against them	Governments, businesses	Disrupt financial website
State actors	Spy on citizens, disrupt foreign government	Users, governments	Read citizen's email messages

Two Rights & A Wrong

1. Script kiddies are responsible for the class of attacks called Advanced Persistent Threats.
2. Hactivists are strongly motivated by ideology.
3. Brokers sell their knowledge of a weakness to other attackers or a government.

See Appendix A for the answer.

Building a Comprehensive Security Strategy

What would a practical, comprehensive security strategy look like? There are five key elements to creating a practical security strategy: block attacks, update defenses, minimize losses, use layers, and stay alert. These elements are by no means new: these tactics go back to the days of medieval castles in Europe and probably much earlier. Understanding these key elements as they were used during the Middle Ages helps bring them into focus for developing practical security today.

Block Attacks

The word *castle* comes from a Latin word meaning *fortress*, and most ancient castles served in this capacity. One of a castle's primary functions was to protect the king's family and citizens of the countryside in the event of an attack from an enemy.

A castle was designed to block enemy attacks in two distinct ways. First, a castle was surrounded by a deep moat that was filled with water, which prevented the enemy from getting close to the castle. In addition, many castles had a high protective stone wall between the moat and the outer walls of the castle. The purpose of the moat and protective wall was to create a *security perimeter* around the castle: any attacker would have to get through the strong perimeter to get inside.

Effective cybersecurity follows this same model of blocking attacks by having a strong security perimeter. Usually, this security perimeter is part of the computer network to which a personal computer is attached, and in most settings like a school or business, it is maintained by security professionals. If attacks are blocked by the network security perimeter, the attacker will be unable to reach the personal computer on which the information is stored. Security devices can be added to a computer network that will continually analyze traffic coming into the network from the outside (such as email or webpages) and block unauthorized or malicious traffic.

In addition to perimeter security, most castles provided *local security*. If an arrow shot by an attacker traveled over the moat and outer wall, those inside the castle could be vulnerable to these attacks, even if there was a strong security perimeter. The solution was to provide each defender with a personal shield to deflect the arrows. This analogy also applies to cybersecurity. As important as a strong network security perimeter is to blocking attacks, some attacks will slip through the defenses. It is vital to also have local security on the devices connected to the network to defend against any attack that breaches the perimeter.

Update Defenses

Suppose a castle came under attack by an enemy, and the defenders started preparing for any arrows that may be shot over the moat and outer wall. The defenders knew from past experiences that these arrows had stone arrowheads, and they were preparing their defenses appropriately for that type of attack.

But suddenly arrows started raining down that had metal arrowheads, something that they had never experienced before. And then some arrows come in that had their tips on fire. Imagine the panic of the defenders trying to protect themselves and their property from this previously unknown new attack. The defenders must quickly adapt to these new attack techniques by continually updating their own defenses.

Today's cyberattackers are equally, if not more, inventive than attackers of 1,000 years ago. New types of attacks appear around the clock. It is essential that cybersecurity defenses be continually updated to protect against the latest attacks, simply because what worked yesterday may not work today in warding off an attack. Updating defenses typically involves applying the latest updates sent from vendors to protect software and hardware. It may even require regularly checking vendor websites to determine if an update has become available but is not widely publicized.

Minimize Losses

As a flaming arrow sails over the castle wall, it might strike a bale of hay and set it ablaze. If the defenders were not prepared with a bucket of water to douse the flames, then the entire castle could burn up. Being prepared to minimize losses was essential in defending a castle.

Likewise, in cybersecurity, it is important to realize that some attacks likely will get through security perimeters and local defenses. It is important that action be taken in advance to minimize losses. This may involve keeping backup copies of important data stored in a safe place, or, for an organization, it may mean having an entire business recovery policy that details what to do in the event of a successful attack.

Use Layers

Despite their massive size, some castles did fall to enemy attacks. Over time castle architects began to modify the design of castles to provide even greater protection. These architects realized that if one defense was breached, then the castle would be overrun. Thus, they started incorporating multiple layers of defense, requiring the enemy to breach not just a single defense but two or three, or even more.

By the late medieval times, a new innovation in castle design became popular. An outer "curtain wall" of stone as much as 6 feet (2 meters) thick was built around the castle walls. This resulted in a castle with two separate walls, essentially producing a "castle within a castle."

Another innovation based on layers was the introduction of a "gatehouse." The entrance door to the castle was considered its weakest point since it had to be constructed of wood that could then be raised and lowered (raising and lowering a heavy stone door would have been impossible). Architects began designing castles with a gatehouse as the entrance to the castle. This was a fortified entryway that was filled with obstacles, such as multiple portcullis gates (heavy vertical gates of a latticed grille) and doors that forced the attackers to breach multiple defenses. And gatehouses also contained "arrow slits" so that defenders could fire arrows at the attackers while they attempted to breach the portcullis gates and even holes in the ceiling through which boiling water could be poured on the enemy!

Like castles, cybersecurity defenses rely heavily on layers to thwart threat actors. Modern-day attackers must then breach multiple defenses—each ideally independent of other defenses—making it significantly more difficult to reach the goal. In fact, using layers can also be used as a discouragement to the attackers to convince them to give up and find an easier target.

Stay Alert

How protected would a castle be if the defenders were asleep or cowered in fear under a bed? It was vital that all those defending the castle would stay alert and be constantly vigilant to join the fight.

The same is true today. It is important to create a "cybersecurity posture" so that security is considered when making decisions. These decisions range from what software to use to what to do when a suspicious email is received.

Cybersecurity cannot be considered the task of "somebody else" but is instead the responsibility of all users. This involves knowing what to do, the skills to take the necessary steps, and the proper motivation to stay secure.

Two Rights & A Wrong

1. Using layers can be a discouragement to the attackers to convince them to give up and find an easier target.
2. Updating defenses typically involves applying the latest updates sent from vendors to protect software and hardware.
3. There are four key elements to creating a practical security strategy.

○ See Appendix A for the answer.

Module Summary

- Preventing attacks is very hard. There are several reasons why it is difficult to defend against today's attacks. One reason is that virtually all devices are connected to the Internet. Other reasons include the speed of the attacks, greater sophistication of attacks, the availability and simplicity of attack tools, faster detection of weaknesses by attackers, delays in security updating, weak security update distribution, distributed attacks coming from multiple sources, and user confusion.
- Security means being free from care. Sometimes security is defined as the state of being free from danger, which is the goal of security, while other times it is defined as the measures taken to ensure safety, which is the process of security. Since complete security can never be fully achieved, the focus of security is more often on the process instead of the goal. There is an important relationship between security and convenience. This relationship between these two is known as inversely proportional: as security is increased, convenience is decreased. The more secure something becomes, the less convenient it may become to use.
- Several terms are used when describing security for technology: computer security, information security, IT security, and information assurance, to name just a few. Whereas each has its share of proponents and slight variations of meanings, the term cybersecurity may be the most relevant. Cybersecurity has been called both an art and a practice. An art is the systematic application of knowledge and skills, while a practice is a repeated or continual action. Cybersecurity should protect the confidentiality, integrity, and availability of information that provides value to people and enterprises. It should also ensure authentication, authorization, and accounting. Cybersecurity may be defined as that which protects the integrity, confidentiality, and

availability of information on the devices that store, manipulate, and transmit the information through products, people, and procedures.

- As with many advanced subjects, cybersecurity has its own set of terminology. The terminology often used with cybersecurity includes asset, threat, threat agent, vulnerability, attack vector, threat likelihood, and risk.
- There are several reasons why cybersecurity is important. Cybersecurity can help prevent data theft. A business must guard against data theft because attackers are eager to steal proprietary business information, such as research for a new product or a list of customers. Individuals are also often the targets of data thievery. Identity theft involves stealing another person's personal information and then using the information to impersonate the victim, often for financial gain. Cybersecurity can also help foil cyberterrorism, or attacks are intended to cause panic or provoke violence among citizens. There are several international, national, and state laws enacted to protect the privacy of electronic data. Businesses that fail to protect data they possess from cybercriminals may face both legal actions as well as significant financial penalties. For businesses, cleaning up after an attack by a business diverts time, money, and other resources away from normal activities. Cybersecurity can help businesses maintain productivity.
- The types of criminals behind attacks fall into several categories. The term "cybercriminals" generally refers to someone who attacks for financial gain. Script kiddies do their work by downloading automated attack software from websites and then use it to attack systems. A broker is an attacker who sells his knowledge of a vulnerability to others. One of the largest information security threats to a business actually comes from insiders who are employed there. Cyberterrorists are motivated by their principles and beliefs and turn their attacks on the network and computer infrastructure to cause panic among citizens. Hactivists use attacks as a means of protest or to promote a political agenda. State actors have been funded by government agencies to attack foreign governments and even their own citizens who they consider hostile or threatening.
- A practical, comprehensive security strategy involves five key elements. The first is to block attacks by having a strong security perimeter, both on the network and on the personal computer as well. Another strategy is to regularly update defenses to protect against the latest attacks. Also, it is important to minimize losses for any attacks that may be successful. Using layers of defenses can also prevent successful attacks. Finally, it is vital to constantly stay alert to attacks.

Key Terms

accounting
Advanced Persistent Threat (APT)
asset
attack vector
authentication
authorization
availability
broker
confidentiality
cybercriminal
cybersecurity
cyberterrorism
cyberterrorist
General Data Protection Regulation (GDPR)
Gramm-Leach-Bliley Act (GLBA)
hactivist
hacker
Health Insurance Portability and Accountability Act (HIPAA)
identity theft
insider
integrity
Payment Card Industry Data Security Standard (PCI DSS)
risk
Sarbanes-Oxley Act (Sarbox)
script kiddie
state actor
threat
threat actor
threat agent
threat likelihood
vulnerability

Review Questions

1. Which of the following is NOT a reason why it is difficult to defend against today's attackers?
   a. Faster detection of vulnerabilities
   b. Complexity of attack tools
   c. Weak security update distribution
   d. Greater sophistication of attacks

2. Which of the following accounts for the greatest difficulty in preventing attacks?
   a. Availability and simplicity of attack tools
   b. Delays in security updating
   c. Distributed attacks
   d. User confusion

3. In a general sense, what is security?
- **a.** It is only available on specialized computers.
- **b.** It is protection from only direct actions.
- **c.** It is the steps necessary to protect a person or property from harm.
- **d.** It is both an art and a science.

4. Which of the following ensures that only authorized parties can view information?
- **a.** Confidentiality
- **b.** Authorization
- **c.** Integrity
- **d.** Availability

5. Why can brokers command such a high price for what they sell?
- **a.** Brokers are licensed professionals.
- **b.** The attack targets are always wealthy corporations.
- **c.** The vulnerability they uncover was previously unknown and is unlikely to be patched quickly.
- **d.** Brokers work in teams and all the members must be compensated.

6. Which of the following is NOT a successive layer in which information security is achieved?
- **a.** Products
- **b.** People
- **c.** Policies and procedures
- **d.** Purposes

7. What is a class of attacks by state actors that use innovative attack tools to silently extract data over an extended period of time?
- **a.** RPP
- **b.** XLX
- **c.** APT
- **d.** GOR

8. What is a person or element that has the power to carry out a threat?
- **a.** Threat actor
- **b.** Agent
- **c.** Risk exploiter
- **d.** Cyber invader

9. In cybersecurity, what is a flaw or weakness that allows an attacker to bypass security protections?
- **a.** Access
- **b.** Vulnerability
- **c.** Worm hole
- **d.** Access control

10. Which of the following ensures that individuals are who they claim to be?
- **a.** Demonstration
- **b.** Authentication
- **c.** Accounting
- **d.** Certification

11. Which of the following requires that enterprises must guard protected health information and implement policies and procedures to safeguard it?
- **a.** Hospital Protection and Insurance Association Agreement (HPIAA)
- **b.** Sarbanes-Oxley Act (Sarbox)
- **c.** Gramm-Leach-Bliley Act (GLBA)
- **d.** Health Insurance Portability and Accountability Act (HIPAA)

12. Which of the following is motivated for the sake of their principles or beliefs?
- **a.** Cyberterrorists
- **b.** Insiders
- **c.** Script kiddies
- **d.** Computer spies

13. What is the difference between a hactivist and a cyberterrorist?
- **a.** A hactivist is motivated by ideology while a cyberterrorist is not.
- **b.** Cyberterrorists always work in groups while hactivists work alone.
- **c.** The aim of a hactivist is not to incite panic like cyberterrorists.
- **d.** Cyberterrorists are better funded than hactivists.

14. Lorenzo has decided to make regular backup copies of information from his laptop and store it in a safe place. Which of the following principles is Lorenzo following?
- **a.** Minimizing losses
- **b.** Blocking attacks
- **c.** Updating defenses
- **d.** Using layers

15. Which of the following is NOT classified as an insider?
- **a.** Business partners
- **b.** Contractors
- **c.** Cybercriminals
- **d.** Employees

16. What is an objective of state actor?
- **a.** To right a perceived wrong
- **b.** To spy on citizens
- **c.** To sell vulnerabilities to the highest bidder
- **d.** To earn fortune over fame

17. Which of the following requires banks and financial institutions to protect all electronic and paper containing personally identifiable financial information?
- **a.** California Savings and Loan Security Act (CS&LSA)
- **b.** Sarbanes-Oxley Act (Sarbox)

c. Gramm-Leach-Bliley Act (GLBA)
d. USA Patriot Act

18. Which of the following ensures that the information is correct and no unauthorized person or malicious software has altered that data?
a. Integrity
b. Obscurity
c. Layering
d. Confidentiality

19. Which of the following is a type of action that has the potential to cause harm?
a. Hazard
b. Risk
c. Threat
d. Peril

20. Bella is explaining to her friend that her new role at work requires her to block the pathways for an attack. Which of the following terms would Bella use to explain what this pathway is?
a. Interception
b. Attack vector
c. Cybersecurity intrusion
d. Asset roadway

Hands-On Projects

Project 1-1: Examine Data Breaches – Visual

In this project, you view the biggest data breaches resulting in stolen information through a visual format.

1. Open your web browser and enter the URL **https://www.informationisbeautiful.net/visualizations/worlds-biggest-data-breaches-hacks/**. (If you are no longer able to access the site through this web address, use a search engine to search for "Information Is Beautiful World's Biggest Data Breaches.")
2. This site will display a visual graphic of the data breaches, similar to Figure 1-6.

Figure 1-6 World's Biggest Data Breaches & Hacks webpage

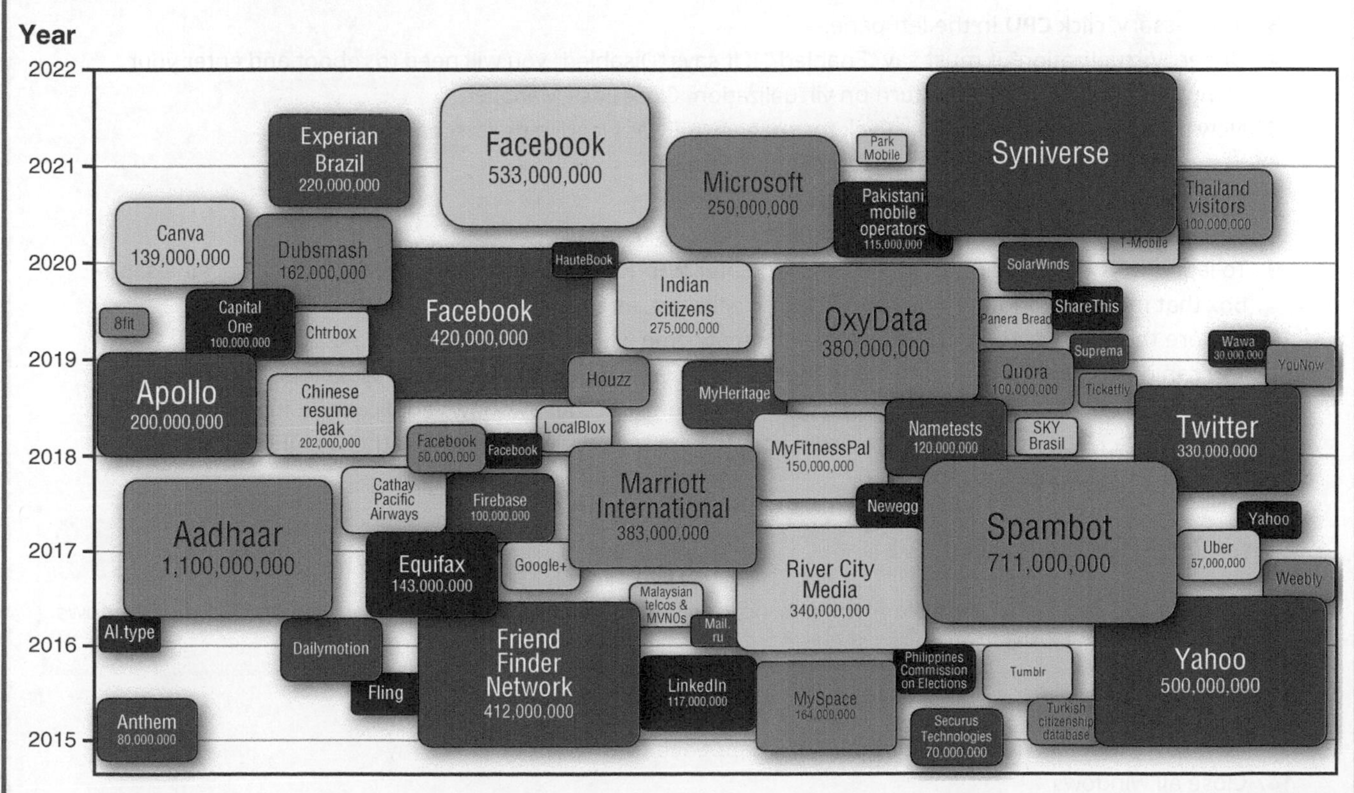

Continued

3. Scroll down the page to view the data breaches by year. Note that the size of the breach is indicated by the size of the bubble.
4. Scroll back up to the top.
5. Hover over several of the bubbles to read a quick story of the breach.
6. Note the color of the bubbles that have an "Interesting Story." Click one of the bubbles and read the story. When finished, close only this tab in your browser.
7. Click **Filter** to display the filter menu.
8. Under Sector, click **retail** to view those breaches related to the retail industry.
9. Click one of the bubbles and read the story.
10. Click **Reset** in the filter menu.
11. Select the sector **finance**.
12. Select the method **poor security**.
13. Click one of the bubbles and read the story.
14. Create your own filters to view different types of breaches. Does this graphic convey a compelling story of data breaches?
15. How does this visualization help you with the understanding of threats?
16. Close all windows.

Project 1-2: Configure Microsoft Windows Sandbox

A *sandbox* is an isolated "virtual" computer (called a "virtual machine") within a "physical" computer. Anything done within a sandbox will impact only this virtual machine and not the underlying computer. And once you close the sandbox then nothing remains on your computer; when you launch the sandbox again it is just like starting over again. A sandbox is an ideal tool for downloading software and testing it to be sure that it contains no malware without impacting the physical computer. In this project, you will configure the Microsoft Windows 11 Sandbox.

1. First check if your system has virtualization turned on. Right-click the Start icon on the taskbar and select **Task Manager**.
2. Click the **Performance** tab.
3. If necessary, click **CPU** in the left pane.
4. Under "Virtualization" it must say "Enabled." If it says "Disabled" you will need to reboot and enter your computer's BIOS or UEFI and turn on virtualization. Close Task Manager.
5. Now enable Windows Sandbox. Click the magnifying glass icon in the taskbar.
6. In the Windows search box, enter **Windows Features**.
7. Click **Turn Windows features on or off**.
8. Click the **Windows Sandbox** check box to turn on this feature.
9. To launch Windows Sandbox, click **Start** and enter **Windows Sandbox**. A protected virtual machine sandbox that looks like another Windows instance will start, as shown in Figure 1-7.
10. Explore the settings and default applications that come with the Windows Sandbox.
11. You can download a program through the Microsoft Edge application in Windows Sandbox. (Edge is included within Windows Sandbox along with a handful of other Windows applications, including access to OneDrive.) Open Edge and go to **www.google.com** to download and install the Google Chrome browser in the Windows Sandbox.

Note 15

You can also copy an executable file from your normal Windows environment and then paste it to the Windows Sandbox desktop to launch it.

12. After the installation is complete, close the Windows Sandbox.
13. Now relaunch the Windows Sandbox. What happened to Google Chrome? Why?
14. Close all windows.

Figure 1-7 Windows Sandbox

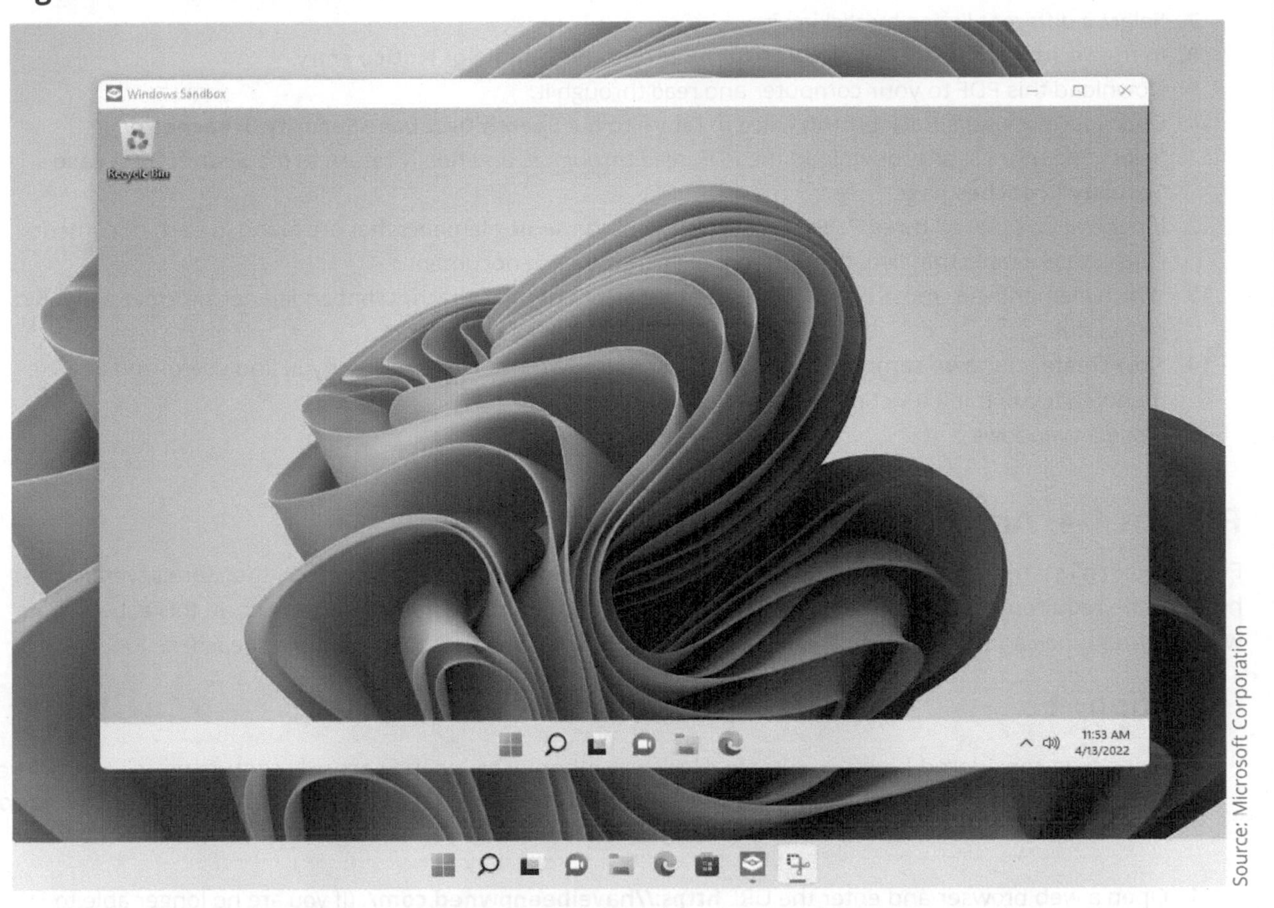

Note 16

After configuring the Windows Sandbox, you can use it for the projects in this book if you do not want to download and install software on your physical computer.

Project 1-3: Comparing Data Breach Notification Letters

All states have their own laws requiring that victims be notified if their personal information has been stolen. However, these notifications have come under criticism for not containing useful information for victims. Many businesses are wary of bad press and legal liability and are reluctant to provide more information than is required. Also, breach notification requirements vary from industry to industry and state to state. Finally, new privacy laws are setting shorter and shorter deadlines for breach notification, and sometimes new information comes to light weeks or even months after a breach is first discovered. In this project, you will compare several California data breach notifications.

1. Open a web browser and enter the URL **https://oag.ca.gov/privacy/databreach/list**. (If you are no longer able to access the site through this web address, open a search engine and enter "California search database security breaches.")
2. Scroll down to see a listing of the most recent California data breaches. Does the number surprise you?
3. Select one of the first listings by clicking it.
4. In the **Submitted Brief Notification Sample**, click the **Sample of Notice** entry.
5. Download this PDF to your computer and read through it.
6. Click your browser's Back button twice to return to the **Search Database Security Breaches** page.

Continued

7. Select a different listing by clicking it.
8. In the **Submitted Brief Notification Sample**, click the **Sample of Notice** entry.
9. Download this PDF to your computer and read through it.
10. Click your browser's Back button twice to return to the **Search Database Security Breaches** page.
11. Select one more listing, download the PDF, read through it, and finally return to the **Search Database Security Breaches** page.
12. If necessary, open all three PDFs. Create a table of the same elements that are found in each document. Also create a table that lists the different elements in each document.
13. Which elements are most useful if you were the victim of this breach? What additional information would be helpful?
14. Now create your own sample notification letter. Include those elements that you find useful and add elements that you think a victim would benefit from.
15. Close all windows.

Project 1-4: Are You a Victim?

Even though all states require some type of notification sent to victims of a data breach, there are several loopholes in the requirements and not all users pay strict attention to these notification emails. In this activity, you will test your email addresses to determine if they are contained in a database of known breaches.

Caution !

This website is considered highly reputable. However, other websites may actually capture your email address that you enter and then sell it to marketers as a valid email address. You should be cautious about entering your email address in a site that does not have a strong reputation.

1. Open a web browser and enter the URL **https://haveibeenpwned.com/**. (If you are no longer able to access the site through this web address, open a search engine and enter "Have I been pwned.")
2. Scroll down and note the **Largest breaches**. Also, note the total number of **pwned accounts**.
3. Enter one of your email addresses in the box and click **pwned?**
4. If this email address has been stolen and listed in the database, you will receive a **Oh no – pwned!** message. If this email address has not been stolen, enter another of your email addresses.
5. Scroll down to **Breaches you were pwned in**.
6. Read the information about the breach, and particularly note the **Compromised data** of each breach. Do you remember being alerted to these data breaches with a notification letter?
7. For any breaches that list **Passwords** in the **Compromised data**, this serves as a red flag that your password for this account was also stolen. Although the stolen password should be "scrambled" in such a way that an attacker would not be able to view it, that may not always be the case. You should stop immediately and change your password at once for that website.

Note 17

Other information listed as compromised data, while important, may be difficult or impossible to change, such as a phone number or physical address. The most critical item that can be changed and should be changed are any passwords.

8. Enter another email address and looked for **Compromised data** that shows any exposed passwords. Change the passwords for those accounts as well.
9. What are your feelings now that you know about your compromised data? Does this inspire you to take even greater security protections?
10. Close all windows.

Case Projects

Case Project 1-1: Personal Attack Experiences

What type of computer attack have you (or a friend or another student) experienced? When did it happen? What type of computer or device was involved? What type of damage did it inflict? What had to be done to clean up following the attack? How was the computer fixed after the attack? What could have prevented it? List the reason or reasons you think that the attack was successful. Write a one-page paper about these experiences.

Case Project 1-2: Security Podcasts or Video Series

Many different security vendors and security researchers now post weekly audio podcasts or video series on YouTube on security topics. Locate two different podcasts and two different video series about computer security. Listen and view one episode of each. Then, write a summary of what was discussed and a critique of the podcasts and videos. Were they beneficial to you? Were they accurate? Would you recommend them to someone else? Write a one-page paper on your research.

Case Project 1-3: Sources of Security Information

The following is a partial overall list of some of the sources for security information:

- Security content (online or printed articles that deal specifically with unbiased security content)
- Consumer content (general consumer-based magazines or broadcasts not devoted to security but occasionally carry end-user security tips)
- Vendor content (material from security vendors who sell security services, hardware, or software)
- Security experts (IT staff recommendations or newsletters)
- Direct instruction (college classes or a workshop conducted by a local computer vendor)
- Friends and family
- Personal experience

Create a table with each of these sources and columns listing Advantages, Disadvantages, Example, and Rating. Use the Internet to complete the entire table. The Rating column is a listing from 1–7 (with 1 being the highest) of how useful each of these sources is in your opinion. Compare your table with other learners.

References

1. Vigna, Paul, "Hackers Steal $540 Million in Crypto From 'Axie Infinity' Game," *The Wall Street Journal*, Mar. 29, 2022, accessed Apr. 8, 2022, https://www.wsj.com/articles/hackers-steal-540-million-in-crypto-from-axie-infinity-game-11648585535.
2. Davis, Jessica, "Alleging security failures caused data breach, patients sue Montana's Logan Health," *SC Magazine*, Apr. 6, 2022, accessed Apr. 8, 2022, https://www.scmagazine.com/analysis/breach/alleging-security-failures-caused-data-breach-patients-sue-montanas-logan-health.
3. "The Works forced to shut some shops after cyber-attack," *BBC News*, Apr. 5, 2022, accessed Apr. 8, 2022, https://www.bbc.com/news/business-60993635.
4. Martin, Andrew, "Mailchimp says it was breached and user accounts accessed," *Bloomberg*, Apr. 4, 2022, accessed Apr. 8, 2022, https://www.bloomberg.com/news/articles/2022-04-04/mailchimp-says-it-was-breached-and-user-accounts-accessed.
5. "Malware," *AV-Test*, accessed Apr. 8, 2022, https://www.av-test.org/en/statistics/malware/.
6. Kress, Robert, "How to develop a cyber-competent boardroom," *Accenture*, Jan. 5, 2022, accessed Apr. 8, 2022, https://www.accenture.com/us-en/blogs/security/cyber-competent-boardroom.

7. "94% of Malware is delivered via Email," *ClearTech* Group, accessed Apr. 16, 2022, https://www.cleartechgroup.com/94-of-malware-is-delivered-via-email/#:~:text=After%20reviewing%20real%2Dworld%20data,compromise%20business%20networks%20and%20devices.
8. Bekker, Eugene, "What are your odds of getting your identity stolen?" *IdentityForce*, Apr. 15, 2021, accessed Apr. 10, 2022, https://www.identityforce.com/blog/identity-theft-odds-identity-theft-statistics.
9. Coleman, Kevin, "Cyber terrorism now at the top of the list of security concerns," *Defensetech*, accessed Apr. 15, 2022, https://www.military.com/defensetech/2011/09/12/cyber-terrorism-now-at-the-top-of-the-list-of-security-concerns.
10. Sheldon, Robert & Hannah, Katie, "Cyberterrorism," *TechTarget*, retrieved Apr. 10, 2022, https://www.techtarget.com/searchsecurity/definition/cyberterrorism#:~:text=Cyberterrorism%20is%20often%20defined%20as,fear%20in%20the%20target%20population.
11. "Google and Alphabet Vulnerability Reward Program (VRP) Rules," https://bughunters.google.com/about/rules/6625378258649088/google-and-alphabet-vulnerability-reward-program-vrp-rules.
12. "Apple Security Bounty," https://developer.apple.com/security-bounty/.
13. "Insider threats are becoming more frequent and most costly: What businesses need to know now," *IDWatchdog*, accessed Apr. 10, 2022, https://www.idwatchdog.com/insider-threats-and-data-breaches/#:~:text=60%25%20of%20Data%20Breaches%20Are%20Caused%20By%20Insider%20Threats&text=A%20recent%20study%20has%20revealed,in%20the%20same%20time%20period.

Module 2

Personal Cybersecurity

After completing this module, you should be able to do the following:

1. Explain how passwords work and the attacks against them.
2. List the different types of attacks using social engineering.
3. Identify social networking risks and identity theft.
4. Explain how to create a defensive stance.
5. Describe personal security defenses.

Cybersecurity Headlines

Recently, there has been a rash of attacks on individuals' retirement accounts. The common thread in these successful attacks has been the compromising of users' passwords.

First, consider what happens if a threat actor were to compromise your credit or debit card number and start making purchases with it. Consumers in the U.S. are protected by the Fair Credit Billing Act (FCBA), which makes the cardholder liable for only $50 in losses on a compromised credit card. With a debit card, that same limit applies—but only if the loss or theft is reported within two business days. If it is not reported within that time, the liability increases up to $500 for 60 days. If the loss is not reported after 60 days, the consumer could be liable for the entire amount. That's why it's always a good idea to use a credit card when shopping online. But recent research has shown that millennials, the largest demographic group in the U.S., still favor using debit cards for purchases.

Most consumers assume that their defined-contribution 401(k) retirement accounts are also protected like credit cards. After all, retirement accounts contain hundreds of thousands of dollars and represent the financial means by which a retiree will live out the remaining years of their life. Because of this, many believe that these accounts have some of the strongest government regulations and a mechanism for reimbursing consumers if their money is stolen.

But such is not the case.

There is no federal regulation that comprehensively governs cybersecurity protections for retirement plans or plan service providers. The Pension Research Council says that the Employee Retirement Income Security Act (ERISA) of 1974 "is silent on data protection in the form of electronic records, and the U.S. courts have not yet decided whether managing cybersecurity risk is a fiduciary function."[1] In other words, there has been no high-level court decision on whether the financial institutions that manage the funds are responsible in the event of a successful cybersecurity breach. So, with no federal regulations and with no court decisions establishing who is responsible, a retirement account is vulnerable.

In several recent incidents, funds have been stolen from victims' retirement accounts by cybersecurity threat actors due to password attacks. In a suit filed by Heide B., she claimed that her former employer (Abbott Laboratories) and the 401(k)-plan recordkeeper (Alight Solutions) violated ERISA, which governs 401(k) plans, by not protecting the money in her retirement account. Heide had received a written letter from Abbott that her 401(k)-account password had been changed and a $245,000 distribution from her account balance of $362,000 (68 percent of the total) was sent to another bank account, which had long since been emptied of the transferred funds.

Continued

Heide sued Abbott as the plan administrator and sponsor and also sued Alight as recordkeeper. According to Heide, the threat actor changed her account password by using the "forgot password" option. A one-time code was sent to her email address on record, but Heide says she never received it. Because the thief had most likely already compromised her email account, it was easy for the thief to confirm the password reset and then change the password on her retirement account before deleting the email. With access to the retirement account with the new password, the attacker evidently also changed the preferences on her account, including the contact telephone number on file, allowing the thief to successfully impersonate her in calls received from the plan's call center. The attacker also changed the preferred contact method from email to postal "snail" mail to give them more time to disappear with the money before Heide was notified. Heide eventually received a portion of the money stolen ($108,000).

In her suit, the court ruled that Abbott had not committed a "breach of duty" and the claims were dismissed against it. However, while some of the claims against Alight were also dismissed, one was not. That case is continuing.

Another theft of $124,105 occurred to a 401(k) account held by Raymond M. In this instance, the attacker contacted American Trust, who held the retirement account, and requested distribution paperwork from Raymond's plan account. This paperwork was sent to an email address or physical address that the suit claims was "not on record" by American Trust as being one of Raymond's addresses. Again, it is likely that the attacker compromised Raymond's American Trust email account password, reset that password, and then changed the personal preferences including the contact telephone number. After American Trust received the distribution paperwork, it called the phone number on record to verify the transaction, and likely ended up talking to the thief for approval.

There are some interesting twists to this second incident regarding Raymond. First, a voided check had to be sent to American Trust along with the paperwork requesting the funds so that American Trust would have the correct bank and account number of where to send the funds. However, the number on the voided check was only 100. This should have raised a red flag immediately: a check with #100 obviously indicates this bank account was brand new and not the account of 60-year-old Raymond. This was actually a bank account that had just been opened by the attacker under Raymond's name. The second interesting twist was the signature on the paperwork was not handwritten; rather, it was an electronic signature from Adobe Acrobat's "Add Signature" feature.

Financial institutions today are no longer routinely restoring all stolen funds, as Heide learned in the first example. Instead, these financial institutions are inquiring what type of security the victim used in protecting their accounts. A weak password or one that is not unique is an indication (in the financial institution's eyes) that the victim was not concerned about securing their accounts. However, if the victim can show that they routinely use a password manager to store a strong password and have requested additional protections from the retirement plan administrator to prevent material from being sent to the preferences on file, these can go a long way to prove the customer is concerned with security and has done all they can to keep their account secure. This can convince the financial institution to restore a larger portion of stolen funds in the event of an attack on passwords.

Many attacks are directed at specific types of technology, like a laptop computer or a smartphone. However, other attacks are broader in scope and can apply across multiple devices and technologies. These attacks are sometimes called platform-independent attacks, agnostic attacks, or personal cybersecurity attacks since they are directed more toward the person rather than a specific device.

This module explores personal security attacks. It also examines defenses for these attacks.

Personal Security Attacks

Several types of attacks are considered personal security attacks. These include password attacks, attacks using social engineering, attacks targeting social networking, and identity theft.

Password Attacks

Passwords have long been a primary target by threat actors, mainly because users create passwords that are easy to break. Understanding password attacks involves knowing what authentication is and how passwords work, the weaknesses associated with passwords, and how threat actors attack passwords.

What Is Authentication?

Consider this scenario: Riker, Peyton, and Paolo work on a local military base and each afternoon they go to the gym on the base to exercise. As they reach the entrance to the building, each must press their finger on the fingerprint reader to enter the building (a "No Tailgating" policy is strictly enforced). As they walk to the receptionist's desk, Riker holds up his ID card to the card reader so the door to the locker room opens for him. As Peyton searches for his card, the receptionist Li just waves him through to the locker room because she knows him. Riker laughs and says to Li, "It's only because of his flaming red hair that you recognize him, and it runs in Peyton's family!" Paolo, however, is new to the base and must sign in. After Li compares his signature to his membership application on file, she then allows him to enter. In the locker room, each of them opens their locker using a combination lock with a series of numbers that they have memorized.

In this scenario, the three men have been demonstrated to be *genuine* or *authentic*, and not imposters, by the seven elements listed in Table 2-1.

Table 2-1 Elements that prove authenticity

Element	Description	Scenario example
Somewhere you are	Restricted location	Restricted military base
Something you are	Unique biological characteristic that cannot be changed	Fingerprint reader to enter building
Something you have	Possession of an item that nobody else has	Riker's ID card
Someone you know	Validated by another person	Li knows Peyton
Something you exhibit	Genetically determined characteristic	Peyton's red hair
Something you can do	Performance of an activity that cannot be exactly copied	Paolo's signature
Something you know	Knowledge that nobody else possesses	Combination to unlock locker

Because only the real (authentic) person possesses one or more of these elements, they can be considered as types of **authentication**, which is proof of genuineness. These can confirm a person's identity and thus give access to restricted areas or materials while denying access by an imposter. In information technology (IT), these elements are known as *authentication credentials* and are presented to an IT system to verify the genuineness of the user. Although any of these elements can be used as an authentication credential, in most instances, IT authentication uses something you know, something you have, or something you are.

Note 1

Three of these elements (something you know, something you have, and something you are) are called *factors*, while the remaining four (somewhere you are, something you can do, something you exhibit, and someone you know) are called *attributes*. The element *something you exhibit* is often linked to more specialized attributes than the color of hair in the scenario and may even include neurological traits that can be identified by specialized medical equipment.

The most common IT authentication credential is providing information that only the user would know. This is done by presenting a **password**, which is a secret combination of letters, numbers, and/or characters known only by the user.

> **Note 2**
>
> The person credited with inventing the computer password, Fernando "Corby" Corbato, who passed away in late 2019, was a researcher for MIT and worked on the Compatible Time-Sharing System (CTSS), which allowed multiple users to share computer time. He devised a way to isolate users from each other with password-protected user accounts. In his later years, Corbato lamented that passwords had become problematic. He said that the Internet made logins with passwords a "kind of a nightmare."

How Passwords Work

Consider a user who is creating a new account for an online bank. The user first creates (or in some instances is assigned) a username for the account. Then, the user creates a password that satisfies the necessary minimum password requirements set by the bank. The username and password are then transmitted to the bank's server. The server will convert the password to a scrambled set of characters that is unlike the original password. Technically, this scrambling is performed by a *one-way hash algorithm* that creates a scrambled *message digest* (or hash) of the password. Finally, the username and digest (scrambled password) would be stored along with other usernames and digests. Creating a password is illustrated in Figure 2-1.

Figure 2-1 Creating a password

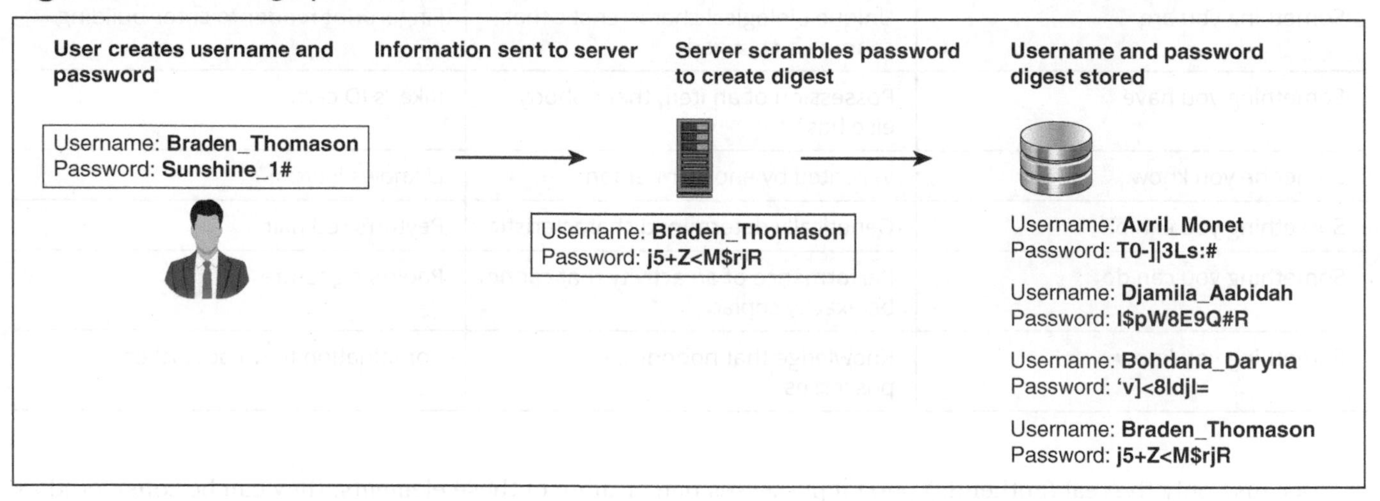

> **Note 3**
>
> The reason why the original unscrambled password is not stored by the server is for security. An attacker who could steal this list of unscrambled passwords would then have unfettered access to all accounts.

The next time that the user logs into his bank account, he would enter both his username and password. This information is again sent to the bank's server, which again creates a message digest on the password just entered by using the same one-way hash used originally when the account was created. The bank's server then looks for a match by comparing this just-created digest with the original stored digest. If the two scrambled digests match, then the user is authenticated, and access is granted. Retrieving a password is illustrated in Figure 2-2.

Figure 2-2 Retrieving a password

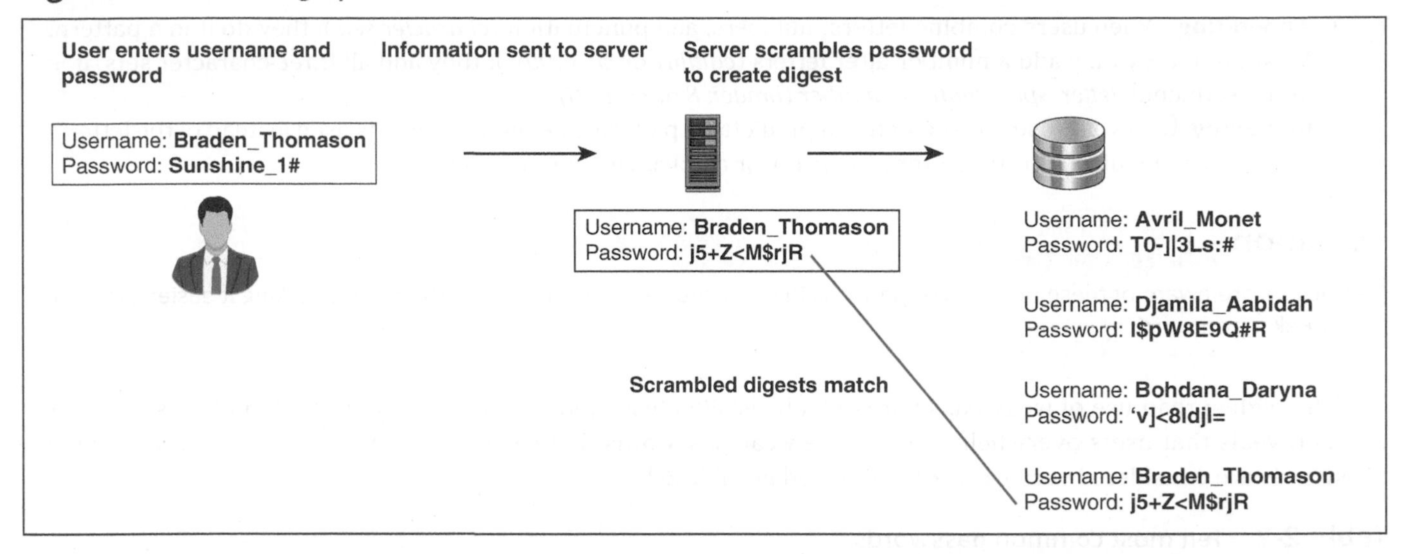

Password Weaknesses

Despite their widespread use, most passwords provide weak protection and are constantly under attack. The weakness of passwords centers on human memory. Human beings can memorize only a limited number of items. Passwords place heavy loads on human memory in multiple ways:

- The most effective passwords are long and complex. However, these are difficult for users to memorize and then accurately recall when needed.
- Users must remember multiple passwords for many different accounts. Most users have accounts for computers and mobile devices at work, school, and home; multiple email accounts; online banking; Internet site accounts; and so on. The average number of accounts per user is estimated to be 207.[2]
- For the highest level of security, each account password should be unique, which further strains human memory.
- Many security policies mandate that passwords expire after a set period of time, such as every 45–60 days, when a new one must be created. Some security policies even prevent a previously used password from being recycled and used again, forcing users to repeatedly memorize new passwords.

Note 4

In recognition of the difficulties surrounding expired passwords, a growing trend has been to drop this policy requirement. In 2019, Microsoft changed its long-held policy and recommended that password expiration be dropped, and in 2017, guidelines released by the National Institute of Standards and Technology (NIST) also recommended that password expiration should no longer be used. However, the Payment Card Industry (PCI) still requires that merchants and other providers change their passwords every 90 days. Some security professionals are calling for a modified password expiration so that the length of the password dictates its expiration. For example, a user who creates a 30-character password would not have to change that password for 2 years, while a password that is 15–25 characters in length would expire annually, and one of fewer than 15 characters would have to be reset every 90 days. One company that tried this approach found that calls to their help desk for password resets declined by 70 percent.

Because of the burdens that passwords place on human memory, users take shortcuts to help them memorize and recall their passwords, resulting in a **weak password** that is easy for an attacker to break. Weak passwords typically use a common word as a password (*princess*), a short password (*desk*), a predictable sequence of characters (*abc123*), or personal information (*Hannah*). Another common shortcut that dramatically weakens passwords is to reuse the same password (or a slight derivation of it) for multiple accounts. Although this makes it easier for the user, it also makes it easier for an attacker who compromises one account to access all other accounts.

Even when users attempt to create stronger passwords, they generally follow predictable patterns:

- **Appending**. When users combine letters, numbers, and punctuation (*character sets*), they do it in a pattern. Most often they only add a number after letters (*caitlin1* or *cheer99*). If they add all three-character sets, it is in the sequence *letters+punctuation+number* (*braden.8* or *chris#6*).
- **Replacing**. Users also use replacements in predictable patterns. Generally, a zero is used instead of the letter *o* (*passw0rd*), the digit *1* for the letter *i* (*denn1s*), or a dollar sign for an *s* (*be$tfriend*).

Caution !

Attackers are aware of these predictable patterns in passwords and can search for them, thus making it easier for them to break a password.

The widespread use of weak passwords can be easily illustrated. An analysis of over 733 million stolen passwords reveals that users overwhelmingly create weak passwords that can easily be broken. The 10 most common passwords are considered very weak and are listed in Table 2-2.[3]

Table 2-2 Ten most common passwords

Rank	Password
1	123456
2	123456789
3	password
4	qwerty
5	12345678
6	1234567890
7	12345
8	111111
9	1234567
10	123123

Note 5

This analysis of the stolen passwords reveals some cultural differences as well. For example, more passwords from Germany are related to the subject of *football* (soccer) while passwords in Russia tend to have more ascending number sequences (*123456*), repeating numbers (*666666*), and duplication of strings (*123123*).

A noted security expert summarized the password problem well by stating:

> The problem is that the average user can't and won't even try to remember complex enough passwords to prevent attacks. As bad as passwords are, users will go out of the way to make it worse. If you ask them to choose a password, they'll choose a lousy one. If you force them to choose a good one, they'll write it [down] and change it back to the password they changed it from the last month. And they'll choose the same password for multiple applications.[4]

Note 6

A recent study looked at users who had been told that the password to their account had been stolen in a data breach. Only one-third of the users then changed their passwords. And the users were in no rush to change their passwords: only 3 percent changed their password within 30 days after the breach, while 12 percent waited between 60 to 90 days. Incredibly, only 14 percent of users changed their password to a *stronger* password; all others created passwords that were actually weaker or the same strength as the stolen password by reusing character sequences from their previous password or creating a new password that was similar to other passwords they use.

Attacks on Passwords

To discover a user's password, attackers use different approaches. These include online brute force attacks, password spraying attacks, and using password collections.

> **Note 7**
>
> There are many other types of attacks on passwords. Typically, attackers use a combination of multiple attacks on passwords.

Online Brute Force Attack In an **online brute force attack**, the same account is continuously attacked (called *pounded*) by entering different passwords. However, attackers rarely use an online brute force attack today because it is impractical. Even at two or three tries per second, it could take thousands of years to guess the right password. In addition, most accounts can be set to disable all logins after a limited number of incorrect attempts (such as five), thus putting an end to the threat.

Password Spraying Attack Another password attack uses a more targeted guessing approach. A **password spraying** attack uses one or a small number of commonly used passwords (*Password1* or *123456*) and then uses this same password when trying to log in to several user accounts. Because this targeted guess is spread across many accounts instead of attempting multiple password variations on a single account, it is much less likely to raise any alarms or lock out the user account from too many failed password attempts. Although password spraying may result in occasional correct guesses, it also has a low rate of success.

Password Collections The most successful approach for breaking passwords involves *using technology for comparisons*. Instead of relying upon random guessing, attackers rely on proven technology to break passwords. Threat actors use sophisticated **password crackers**, which is software specifically designed to break passwords. A password cracker program is illustrated in Figure 2-3.

Password crackers perform comparisons between known digests and a user's unknown password digest. For example, an attacker may create their own set of known digests called *candidates* (*Sunday* = *D4RhQ8b2*, *Monday* = *dXCq2s8d*, etc.) and then compare those with the user's digest. When a match occurs, the attacker then knows the underlying user's password, and the password is said to be "broken."

A watershed moment in password attacks occurred in late 2009. An attacker broke into a server belonging to a developer of several popular social media applications. This server contained more than 32 million user passwords, which were later posted on the Internet. Attackers quickly seized upon this opportunity. This "treasure-trove" collection of passwords gave attackers, for the first time, a large corpus of real-world passwords. Because users repeat their passwords on multiple accounts, attackers could now use these passwords as candidate passwords in their attacks with a high probability of success.

> **Note 8**
>
> These password collections also provided attackers advanced insight into the strategic thinking of how users create passwords. For example, on those occasions when users mix uppercase and lowercase in passwords, users tend to capitalize at the beginning of the password, much like writing a sentence. Likewise, punctuation and numbers are more likely to appear at the end of the password, again mimicking standard sentence writing. And a high percentage of passwords were comprised of a name and date, such as *Braden2008*. Such insights are valuable in crafting attacks, significantly reducing the amount of time needed to break a password.

Since then, using stolen password collections as candidate passwords has become today the foundation of password cracking, and almost all password cracking software tools can accept these stolen "wordlists" as input. Websites host lists of these leaked passwords that attackers can download. These sites also attempt to crack submitted password collections. One website boasts over 1.45 *trillion* cracked password digests.

Figure 2-3 Password cracker program

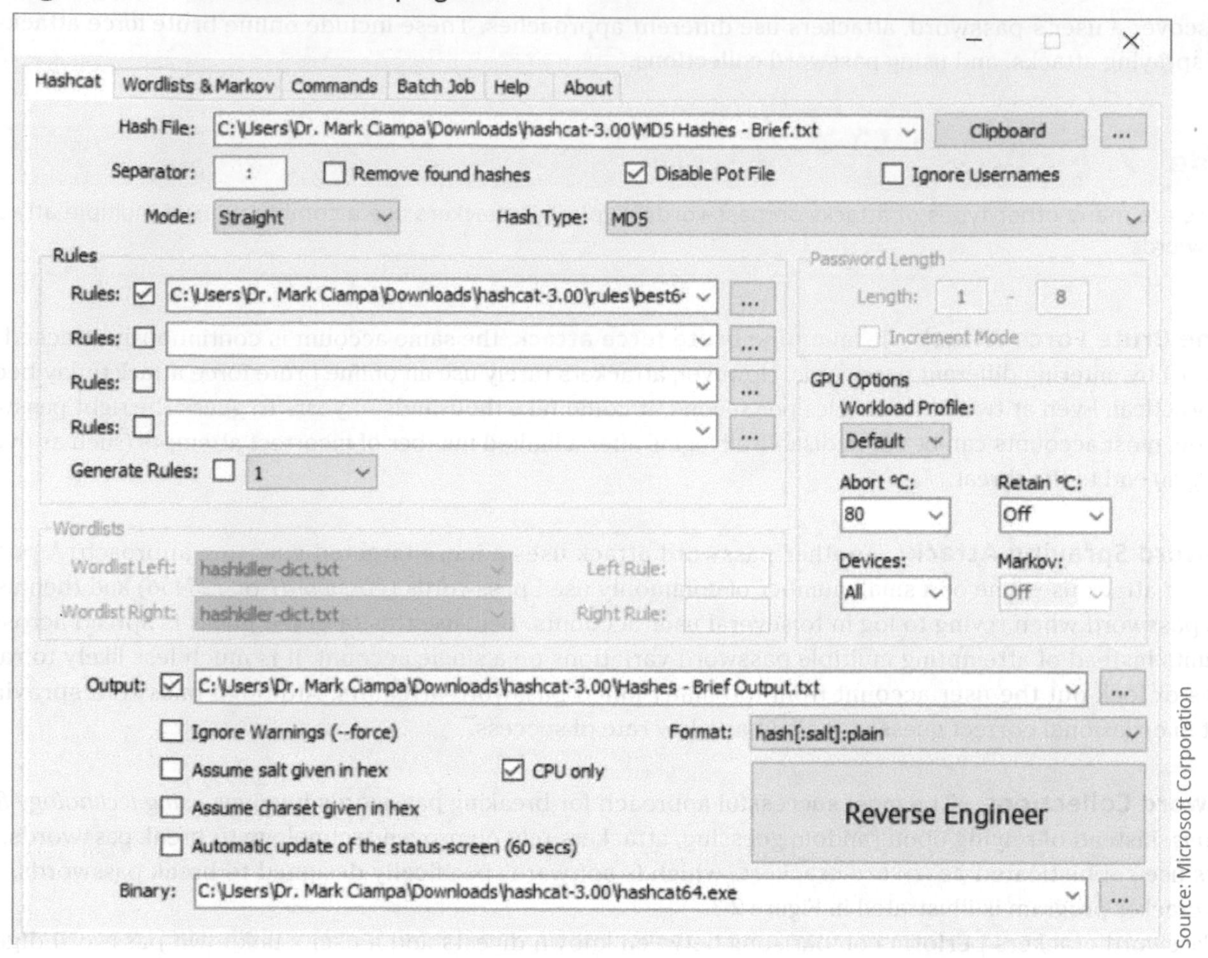

Source: Microsoft Corporation

Attacks Using Social Engineering

Not all attacks rely on technology; in fact, some cyberattacks use little if any technology to achieve their goals. **Social engineering** is a means of using trickery to cause the victim to act in the attacker's favor.

Social engineering relies on an attacker's clever manipulation of human nature to persuade the victim to provide information or take action. Several basic principles make social engineering highly effective. These are listed in Table 2-3 with the example of an attacker pretending to be the chief executive officer (CEO) calling the organization's help desk to have a password reset.

Because many of the psychological approaches involve person-to-person contact, attackers use a variety of techniques to gain trust. For example:

- **Provide a reason**. Many social engineering threat actors are careful to add a reason along with their request. By giving a rationalization and using the word "because," the victim is much more likely to provide the information. For example, *I was asked to call you because the director's office manager is out sick today.*
- **Project confidence**. A threat agent is unlikely to generate suspicion if she enters a restricted area and calmly walks through the building as if she knows exactly where she going (without looking at signs, down hallways, or reading door labels) and even greets people she sees with a friendly *Hi, how are you doing?*
- **Use evasion and diversion**. When challenged, a threat agent might evade a question by giving a vague or irrelevant answer. They could also feign innocence or confusion, or just keep denying any allegations, until the victim eventually believes his suspicions are wrong. Sometimes a threat agent can resort to anger and cause the victim to drop the challenge. *Who are you to ask that? Connect me with your supervisor immediately!*

- **Make them laugh.** Humor is an excellent tool to put people at ease and develop a sense of trust. *I can't believe I left my badge in my office again! You know, some mistakes are too much fun to only make once!*

Table 2-3 Social engineering effectiveness

Principle	Description	Example
Authority	Directed by someone impersonating an authority figure or falsely citing their authority	"I'm the CEO calling."
Intimidation	To frighten and coerce by threat	"If you don't reset my password, I will call your supervisor."
Consensus	Influenced by what others do	"I called last week, and your colleague reset my password."
Scarcity	Something is in short supply	"I can't waste time here."
Urgency	Immediate action is needed	"My meeting with the board starts in 5 minutes."
Familiarity	Victim is well-known and well-received	"I remember reading a good evaluation on you."
Trust	Confidence	"You know who I am."

Social engineering frequently uses **impersonation** (also called identity fraud) or masquerading as a real or fictitious character and then playing out the role of that person on a victim. For example, an attacker could impersonate a help desk support technician who calls the victim, pretends that there is a problem with the network, and asks her for her username and password to reset the account. To impersonate a real person, the threat actor usually must know as much about the person as possible to appear genuine. This type of reconnaissance is known as *credential harvesting* and is commonly carried out by Internet and social networking searches.

Caution !

Often attackers will impersonate individuals whose roles are authoritative because victims generally resist saying "no" to anyone in power. Users should exercise caution when receiving a phone call or email from these types of individuals asking for something suspicious.

Attacks using social engineering include phishing, typo squatting, and hoaxes.

Phishing

One of the most common attacks based on social engineering is phishing. **Phishing** is sending an email or displaying a web announcement that falsely claims to be from a legitimate enterprise in an attempt to trick the user into surrendering private information or taking action.

Note 9

The word *phishing* is a variation on the word "fishing," with the idea being that bait is thrown out knowing that while most will ignore it, some will "bite."

Users are asked to respond to an email or are directed to a website where they are requested to update personal information, such as passwords, credit card numbers, Social Security numbers, bank account numbers, or other information. However, the email or website is actually an imposter and is set up to steal what information the user enters. Or users may receive a fictitious, overdue invoice that demands immediate payment, and in haste, a payment is then made (called an *invoice scam*). Figure 2-4 illustrates an invoice scam phishing email message.

Figure 2-4 Invoice scam phishing email message

INVOICE NUMBER **Product Details**

NOR04142022GB NORTON 360 PROTECTION

Order Summary

INVOICE NO.: NOR04142022GB
Start Date: **2022-04-14**
End Date: **1 year from Start Date**
Payment Mode: Auto debit from account
Status: Completed

--

Product Title	Quantity	Total
NORTON 360 PROTECTION (NOR04142022GB)	1	$566.00 USD
	Sub-total	$566.00 USD
	Discount	00.00
	Total	$566.00 USD

If you wish to stop subscription and ask for a **REFUND** then please feel free to call our Billing Department as soon as possible!
You can Reach us on : **+1 – (877) – (209) – 2392**

Regards,
Refund & Settlement Dept.

Caution !

Note in Figure 2-4 that a telephone number is included that a panicked victim may call to stop the payment. Of course, this telephone number is a direct line to the attackers.

Phishing continues to be a primary weapon used by threat actors and is considered to be one of the largest and most consequential cyber threats facing both businesses and consumers. The APWG (Anti-Phishing Working Group) reported 316,747 phishing attacks in December 2021, the highest monthly total observed since it began its reporting program in 2004. Overall, the number of phishing attacks has tripled since early 2020. The financial sector, which includes banks, is the most frequently attacked industry, accounting for 23.2 percent of all phishing. A growing area is phishing directed against cryptocurrency targets, such as cryptocurrency exchanges and wallet providers.[5]

Caution 

Phishing is also used to validate email addresses. A phishing email can display an image retrieved from a website that is requested when the user opens the email message. A unique code is used to link the image to the recipient's email address, which then tells the phisher that the email address is active and valid. This is the reason most email today does not automatically display images that are received in emails. Users should be cautious in displaying images in emails.

Typo Squatting

If a threat actor cannot trick a user to visit a malicious website through phishing, other tactics are used to redirect the user.

What happens if a user makes a typing error when entering a uniform resource locator (URL) address in a web browser, such as typing *goggle.com* (a misspelling) or *google.net* (incorrect domain) instead of the correct *google.com*? In the past, an error message like *HTTP Error 404 Not Found* would appear. However, today the user is most often directed to a fake look-alike site. These sites may pretend to be the legitimate site and trick the user into entering personal information. Or the site may just be filled with ads for which the attacker receives money for traffic generated to the site. These fake sites exist because attackers purchase the domain names of sites that are spelled similarly to actual sites. This is called **typo squatting**. A well-known site like *google.com* may have to deal with more than 1,000 typo-squatting domains.

Note 10

Enterprises have tried to preempt typo squatting by registering the domain names of close spellings of their website. At one time, top level domains (TLDs) were limited to .com, .org, .net, .int, .edu, .gov, and .mil, so it was fairly easy to register close-sounding domain names. However, today there are over 1200 generic TLDs (gTLDs), such as .museum, .office, .global, and .school. Organizations must now attempt to register a very large number of sites that are a variation of their registered domain name.

Hoaxes

Threat agents can even use social engineering to promote hoaxes. A **hoax** is a false warning, often contained in an email message claiming to come from the IT department. The hoax purports that there is "deadly malware" circulating through the Internet and that the recipient should erase specific files or change security configurations, and then forward the message to other users. However, changing configurations allows an attacker to compromise the system. Or, erasing files may make the computer unstable, prompting the victim to call the telephone number in the hoax email message for help, which is actually the phone number of the attacker.

Social Networking Risks

Social networking is the use of Internet-based social media platforms that allow users to stay connected with friends, family, or peers. Social networking has been called a global revolution, enabling billions of people worldwide to stay in touch with their friends, share experiences, and exchange personal content. It has virtually replaced not only the phone and email for communication but also newspapers and television for current news information. For many users, social networking has become a way of life.

Caution 

Although they are sometimes used interchangeably, social media and social networking are not identical. Social media are forms of electronic communication, such as websites, through which users engage in social networking, which involves creating online communities to share information, ideas, personal messages, videos, and other content.

The very nature of social networking—interacting electronically with a massive base of users whose true identities are hidden—carries a significant risk of becoming a target for attackers. These risks include the following:

- **Personal data can be used maliciously**. Users post personal information on their pages for others to read, such as birthdays, where they live, their plans for the upcoming weekend, and the like. However, attackers can use this information for a variety of malicious purposes. For example, knowing that a person is on vacation could allow a burglar to break into an empty home, the name of a pet could be a weak password that a user has created, or too much personal information could result in identity theft.
- **Users may be too trusting**. Attackers often join a social-networking site and pretend to be part of the network of users. After several days or weeks, users begin to feel they know the attackers and may start to provide personal information or click embedded links provided by the attacker that loads malware onto the user's computer.
- **Social networking security is lax or confusing**. Because social-networking sites by design are intended to share information, these sites have often made it too easy for unauthorized users to view other people's information. To combat this, many sites change their security options on a haphazard basis, making it difficult for users to keep up with the changes.
- **Accepting friends may have unforeseen consequences**. Some social-networking users readily accept any "friend" request they receive, even if they are not familiar with that person. This can result in problems since whoever is accepted as a friend may then be able to see not only all of that user's personal information but also the personal information of their friends.

Note 11

The enormous scope of social networking can be seen by considering just one of the many social networking sites, namely Facebook. Facebook is the world's largest social networking site, with more than two billion people using it every month. This means that almost 37 percent of the world's population are Facebook users.[6]

Identity Theft

Identity theft involves stealing another person's personal information, such as a Social Security number, and then using the information to impersonate the victim, often for financial gain. Using this information to obtain a credit card, set up a cellular telephone account, or even rent an apartment, thieves can make excessive charges in the victim's name. The victim is charged for the purchases and suffers a damaged credit history that can be the cause for being turned down for a new job or denied loans for school, cars, and homes.

Note 12

Identity theft was introduced in Module 1.

The following are some of the actions that have been undertaken by identity thieves:

- Produce counterfeit checks or debit cards and then remove all money from the bank account
- Establish phone or wireless service in the victim's name
- File for bankruptcy under the person's name to avoid eviction
- Go on spending sprees using fraudulently obtained credit and debit card account numbers to buy expensive items such as large-screen televisions that can easily be resold
- Open a bank account in the person's name and write bad checks on that account
- Open a new credit card account, using the name, date of birth, and Social Security number of the identity-theft victim. When the thief does not pay the bills, the delinquent account is reported on the victim's credit report.
- Obtain loans for expensive items such as cars and motorcycles

Two Rights & A Wrong

1. Phishing is sending an email or displaying a web announcement that falsely claims to be from a legitimate enterprise in an attempt to trick the user into surrendering private information or taking action.
2. The most common IT authentication credential is providing information that only the user would know, namely a password.
3. The most effective password attack is a password spraying attack.

See Appendix A for the answer.

Creating a Defensive Stance

The topic of protecting yourself from cyberattacks often leads users to one of two extremes. The first extreme is to think that there is little to do that is truly effective: *There's nothing I can do to protect myself, so why even try?* The second extreme is to take exceptional measures to attempt to create an impregnable defense: *I will never use the Internet again in order to prevent all attacks.*

As is often the case with extremes, both of these are overreactions.

While the number of successful attacks may seem overwhelming to the point that no user can defend herself, that is not the case. There are basic steps that all users can do for their protection. But these actions do not require an impregnable defense.

The starting point is to first create a defensive stance (a *stance* is an attitude, posture, or bearing). And a defensive stance begins with this basic goal: the key to protecting my digital life is to make it *too difficult for an attacker to upend my safety, financial security, and privacy.*

While it is true that with enough time, resources, and skill, a dedicated attacker can defeat a defense, few if any attackers will want to dedicate all of those resources against a single person. The reason is simply that other victims have fewer (if any) defenses that can much more easily and quickly be breached. Thus, *being more secure than the average user will afford you the basic protections that you need.* This is not to say that these are the *only* protections necessary, but they can become the starting point.

Users must take stock of their current digital life and the protections needed by asking these questions:

- What do I need to protect? (That is, what in my digital life can give away critical information tied to my finances, privacy, and safety?)
- How likely is it that it needs protection? (That is, what is your current personal level of exposure to threats?)
- Will the effort be worth it to protect my digital life? (That is, do you want to spend the energy to protect yourself?)

Fortunately, creating a defensive stance is very doable. The most likely place to start is with personal security defenses, for a breach of these defenses can result in upheaval to your safety, financial security, and privacy.

Note 13

Remember that your cybersecurity defenses do not have to be perfect; they just have to be good.

Two Rights & A Wrong

1. The key to protecting your digital life is to make it too difficult for an attacker to upend your safety, financial security, and privacy.
2. Being more secure than the average user will afford you the basic protections that you need.
3. No matter what defenses you use, they must be perfect to defeat all attackers.

See Appendix A for the answer.

Personal Security Defenses

Despite the growing number of attacks on users' personal security, defenses can be used to ward off these attacks. These defenses include password defenses, recognizing social engineering attacks, taking steps to avoid identity theft, and reducing social networking risks.

Password Defenses

There are several defenses against password attacks. These include creating strong passwords using password managers and other password defenses.

Creating Strong Passwords

Strong passwords should be created for each separate account. Creating strong passwords includes observing the following guidelines:

- Do not use passwords that consist of dictionary words or phonetic words.
- Do not repeat characters (*xxx*) or use sequences (*abc, 123, qwerty*).
- Do not use birthdays, family member names, pet names, addresses, or any personal information.
- Do not use short passwords.

One way to make passwords stronger is to use nonkeyboard characters, or special characters that do not appear on the keyboard. For the Microsoft Windows operating system, these characters are created by holding down the *ALT* key while simultaneously typing a number on the numeric keypad (not the numbers across the top of the keyboard). For example, *ALT* + *0163* produces the £ symbol. A list of all the available nonkeyboard characters can be seen by clicking *Start* and entering *charmap.exe*, and then clicking a character. The code ALT + 0*xxx* will appear in the lower-left corner of the screen (if that character can be reproduced in Windows). Figure 2-5 shows a Windows character map.

Figure 2-5 Windows character map

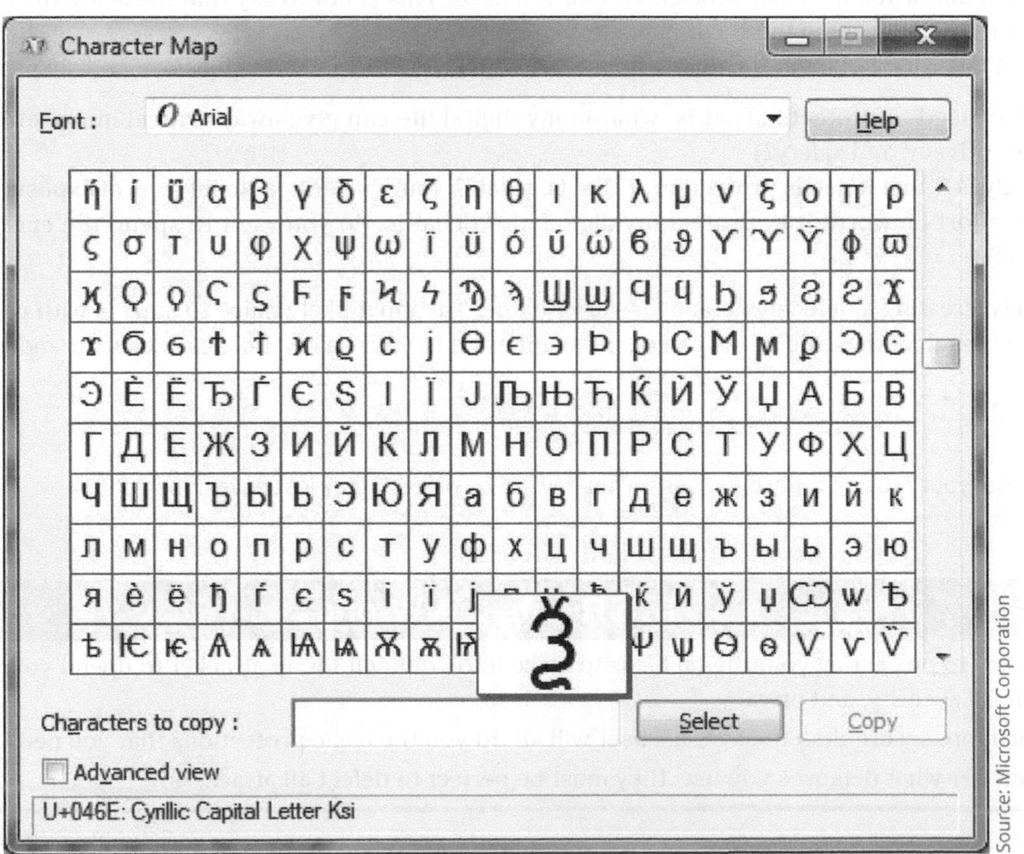

Source: Microsoft Corporation

However, the most critical factor in a strong password is not complexity but length: a longer password is always more secure than a shorter password. This is because the longer a password is, the more attempts an attacker must make to attempt to break it.

Note 14

In technical terms, increasing the length of a password increases the strength *exponentially*, while increasing the complexity only increases it *linearly*.

Users often overestimate the strength of their passwords based on password length. The formula for determining the number of possible passwords from a specific password length is *Number-of-Keyboard-Keys ^ Password-Length = Total-Number-of-Possible-Passwords*. Table 2-4 illustrates the number of possible passwords for different password lengths using a standard 95-key keyboard, along with the average attempts needed to break a password. The average attempts to break a password is calculated as one-half of the total number of possible passwords. (That is because an attack could break the password on the first attempt or the very last attempt.)

Table 2-4 Number of possible passwords

Keyboard keys	Password length	Number of possible passwords	Average attempts to break password
95	2	9025	4513
95	3	857,375	428,688
95	4	81,450,625	40,725,313
95	5	7,737,809,375	3,868,904,688
95	6	735,091,890,625	367,545,945,313

Although passwords that take millions of average attempts to break them may sound strong, in reality, they are not. Consider a computer called Brutalis that is advertised to break passwords. It is said to be "the fastest, meanest, most hardcore password cracker money can buy." Brutalis can generate *8.5 billion candidate passwords per second.* This illustrates that shorter passwords can easily be broken by attackers.

Note 15

Brutalis also comes with a three-year warranty.

Using Password Managers

In addition to the characteristics listed previously regarding weak passwords (such as using a common dictionary word, creating a short password, or using personal information in a password), there are two additional characteristics of weak passwords:

- Any password that can be *memorized* is a weak password.
- Any password that is *repeated* on multiple accounts is a weak password.

Because of the limitations of human memory and the fast-processing speed of today's computers used by attackers, *users can't memorize multiple long passwords that can resist attacks.*

Instead of relying on human memory for passwords, security experts today universally recommend that technology be used instead to create, store, and retrieve passwords. The technology used for securing passwords is called a

password manager. There are different types of password managers, ranging from applications that can be installed on a smartphone or laptop to online "vaults" that store passwords. Each has its own strengths and weaknesses, but all provide a higher degree of password security than can be achieved by relying on human memory (which often results in reusing passwords on multiple accounts or creating passwords that are easy to remember but also easy for attackers to break).

Note 16

Most web browsers allow a user to save a password that has been entered while using the browser. However, this feature has several disadvantages. Users can only retrieve passwords on the computer on which they are stored (unless the browser information is synched with other computers). Also, the passwords may be vulnerable if another user is allowed access to their computer. In addition, applications are freely available that allow all of the passwords to be displayed without entering a master password.

Password managers are not only used to store and retrieve passwords. They also contain a **random password generator** feature that allows the software to create long and unique passwords. A built-in random password generator can create strong random passwords based on different settings like the password manager KeePass password generator shown in Figure 2-6.

Figure 2-6 KeePass random password generator

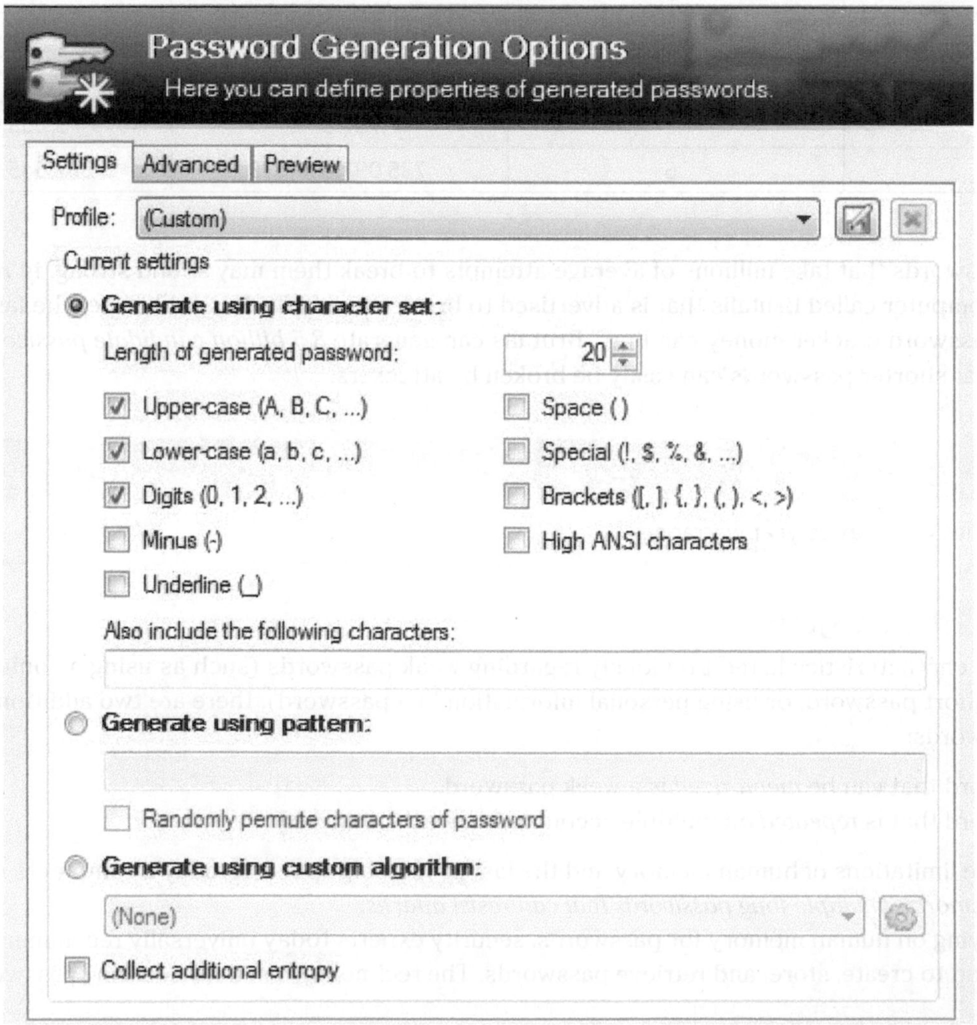

Source: KeePass

The value of using a password manager is that unique strong passwords such as *×oÇ±q¥$Iat=wp¬·hWé2gÑûýÒÄï 8¯çõï|#.×C.75¤íNz²ìù0s4¥ùÄãZo* can be easily created, stored, and retrieved for all accounts that require a password.

Caution !

As an alternative to passwords (something you know), several devices now allow users to present their authentication through *biometrics* (something you are), which may include a fingerprint, retina, and even voice recognition. However, there are several issues surrounding biometrics. Biometric authentication is not foolproof, so genuine users may be rejected while imposters are accepted. Biometric devices can also be "tricked" (security researchers have demonstrated that fingerprints can be collected from water glasses and used to trick fingerprint readers on smartphones), and some experts question the sacrifice of user privacy. Also, if a user's biometric information is stolen, it cannot be reset like a password can and may then be available to attackers forever.

Other Password Defenses

Other password defenses can also be used. These include the following:

- **Two-factor authentication (2FA).** Another type of authentication credential is based on the approved user having a specific item in their possession (something you have). Such items are often used with passwords. Because this involves more than one type of authentication credential—both what a user knows (the password) and what the user has (like a smartphone)—this type of authentication credential is called **two-factor authentication (2FA)**. Using 2FA generally means that after entering a password, a text message with another unique code will be sent to the user's smartphone to then be entered as well. This prevents an attacker who has broken a password from using it since it would also require the attacker to be in the possession of the user's smartphone. Many sites today automatically use 2FA. For those sites in which 2FA is an option, it is a good practice to turn it on.
- **Special email account.** Most accounts allow the user to have an email account associated with it, so that if the password is forgotten an email reset link will be sent to that email account. A secure practice is to have a special email account that is *only* used to receive email reset links. Because this email account is not used for any other purpose, it reduces the risk of an attacker compromising it.
- **Anonymous username.** If the option is available to create a username for an account, it is a secure practice to create an anonymous username that cannot be associated with a specific person. That is, instead of using the username *eleftheria_tims* that can be associated with a person, an anonymous username such as *dQtSWUcQtmNSUeT* could be used instead.
- **Fictitious security answers.** Most accounts ask for the answer to a security question such as *Who was your third-grade teacher?* as another means of security. Instead of accurately answering the question, a fictitious answer should be used instead.

Note 17

Because usernames, passwords, and answers to security questions can be stored in a password manager, there is no need to memorize any of this information.

Recognizing Social Engineering Attacks

Two foundational principles should be recognized when combating social engineering attacks. The first principle is that attacks based on social engineering can come at any time without any warning. The second principle is that the attacker presents herself as someone who can be trusted. This means that both of these principles must be counteracted: users must always be aware and be initially suspicious that virtually any email or electronic correspondence could be a social engineering attack (until the sender can be verified). Table 2-5 lists several social engineering defenses.

Table 2-5 Social engineering defenses

Social engineering example	Secure action	Explanation
An email stating you have won a prize and you must send your bank account number for the money to be deposited.	Recognize scams	Any offer for "easy money" should always be rejected.
A text message that a friend vacationing overseas has lost her purse and needs you to immediately purchase gift cards or a money wire transfer to send funds to a foreign bank account.	Think before you click	Attackers employ a sense of urgency to make you act now and think later, so any highly urgent or high-pressure messages should be rejected until they can be verified through another method of communication different from the message itself (like calling the person if a text message was received).
An email from a company that says your credit card will be charged for a recent purchase, but you did not make the purchase.	Research sources	Always be careful of any unsolicited messages and check the domain links to see if the company is real and if the person sending you the email belongs to the organization.
You receive an email from a friend that has an attachment with the subject line *I can't believe this is a picture of you doing this!*	Never download unexpected file attachments	Always verify through a different channel (phone, text message, etc.) with the sender that the attachment is legitimate, especially if there is a sense of urgency with the message.
A text message asks you to donate to a disaster recovery effort due to a tornado that occurred last night.	Reject requests for help	Perform research into the organization asking for funds.
An email says that you are eligible to apply for federal disaster relief and this company will assist you.	Reject offers of help	Go directly to the primary website that assists and never give personal information through an email to an unknown sender.

Caution !

You should never click a URL link contained in an email message. This is because the link that is displayed (such as *www.amazon.com*) may mask the true link hidden in the message (*such as www.evil.com*).

Avoiding Identity Theft

Identity theft occurs when an attacker uses the personal information of someone else, such as a Social Security number, credit card number, or other identifying information, to impersonate that individual with the intent to commit fraud or other crimes. Avoiding identity theft involves two basic steps. The first step is to deter thieves by safeguarding information. This includes the following:

- Shred financial documents and paperwork that contains personal information before discarding it.
- Do not carry a Social Security number in a wallet.
- Do not provide personal information either over the phone or through an email message.
- Keep personal information in a secure location in a home or apartment.
- Choose to receive electronic notification of statements instead of having them sent through postal mail.

Note 18

The United States Postal Service (USPS) offers a free service called Informed Delivery. You can receive each day an email that contains pictures of all the mail to be delivered that day.

The second step is to monitor financial statements and accounts by doing the following:

- Be alert to signs that may indicate unusual activity in an account, such as a bill that did not arrive at the normal time or a large increase in unsolicited credit cards or account statements.
- Follow up on calls regarding purchases that were not made.
- Review financial and billing statements each month carefully as soon as they arrive.

Caution 

Some local banks offer "shred days" when customers can bring in documents to be shredded and destroyed by a licensed document disposal company. However, do not just drop off your documents but instead watch them being shredded.

Legislation has been passed that is designed to help U.S. users monitor their financial information. The **Fair and Accurate Credit Transactions Act (FACTA) of 2003** contains rules regarding consumer privacy. FACTA grants consumers the right to request one free credit report from each of the three national credit-reporting firms every 12 months. If a consumer finds a problem with her credit report, she must first send a letter to the credit-reporting agency. Under federal law, the agency has 30 days to investigate and respond to the alleged inaccuracy and issue a corrected report. If the claim is upheld, all three credit-reporting agencies must be notified of the inaccuracies, so they can correct their files. If the investigation does not resolve the problem, a statement from the consumer can be placed in the file and any future credit reports.

Note 19

Because a credit report can only be ordered once per year from each of the credit agencies, it is recommended that one report be ordered every four months from one of the three credit agencies. This allows you to view a credit report each quarter without being charged for it.

Reducing Social Networking Risk

Social-networking sites contain a wealth of information for attackers, such as providing information to identity thieves or giving attackers insight into answers to users' security questions that are used for resetting passwords (such as *What is your mother's maiden name?*). With all of this valuable information available, social networking sites should be at the forefront of security today; sadly, that is not always the case. Social-networking sites have a history of providing lax security, not giving users a clear understanding of how security features work, and of changing security options with little or no warning.

Several general defenses can be used for any social networking site. First, users should be cautious about what information is posted on social-networking sites. Posting *I'm going to Florida on Friday for two weeks* could indicate that a home or apartment will be vacant for that time, a tempting invitation for a burglar. Other information posted could later prove embarrassing. Asking questions such as *Would my boss approve?* Or *What would my mother think of this?* before posting may provide an incentive to rethink the material one more time before posting.

Second, users should be cautious regarding who can view their information. Certain types of information could prove to be embarrassing if read by certain parties, such as a prospective employer. Other information should be kept confidential. Users are urged to consider carefully who is accepted as a friend on a social network. Once a person has been accepted as a friend, that person will be able to access any personal information or photographs. Instead, it may be preferable to show "limited friends" a reduced version of a profile, such as casual acquaintances or business associates.

Finally, because available security settings in social networking sites are often updated frequently by the site with little warning, users should pay close attention to information about new or updated security settings. New settings often provide a much higher level of security by allowing the user to fine-tune their account profile options.

Two Rights & A Wrong

1. The most critical factor in a strong password is not length but complexity.
2. Password managers are not only used to store and retrieve passwords, but they also contain a password generator feature.
3. In a social engineering attack, the attacker presents herself as someone who can be trusted.

See Appendix A for the answer.

Module Summary

- Many attacks are directed at specific types of technology, like a laptop computer or a smartphone. However, other attacks are broader in scope and can apply across multiple devices and technologies. These attacks are sometimes personal cybersecurity attacks.
- Authentication is proof of genuineness. Multiple elements can confirm a person's identity and thus give access to restricted areas or materials while at the same time denying access to an imposter. Although any of these elements can be used as an authentication credential, in most instances IT authentication uses something you know, something you have, or something you are. The most common IT authentication credential is providing information that only the user would know. This is done by presenting a password, which is a secret combination of letters, numbers, and/or characters known only by the user.
- When a user first creates a password for an account, it is converted to a scrambled set of characters that is unlike the original password. This scrambling is performed by a one-way hash algorithm that creates a scrambled message digest (or hash) of the password. The username and digest (scrambled password) are stored along with other usernames and digests. The next time that the user logs in, they would enter the password that is again "scrambled" (creates a message digest) and then compared it with the original stored digest. If the two scrambled digests match, then the user is authenticated, and access is granted.
- Despite their widespread use, most passwords provide weak protection and are constantly under attack. The weakness of passwords centers on human memory. Human beings can memorize only a limited number of items. Passwords place heavy loads on human memory in many ways. Because of the burdens that passwords place on human memory, users take shortcuts to help them memorize and recall their passwords, resulting in weak passwords that are easy for attackers to break.
- To discover a user's password, attackers use different approaches. In an online brute force attack, the same account is continuously attacked by entering different passwords. However, an online brute force attack is rarely used today by attackers because it is impractical. Another password attack uses a more targeted guessing approach. A password spraying attack uses one or a small number of commonly used passwords and then uses this same password when trying to log in to several user accounts. Although password spraying may result in occasional correct guesses, it also has a low rate of success. The most successful approach for breaking passwords involves using technology for comparisons. Threat actors use sophisticated password crackers, which is software specifically designed to break passwords through comparisons. Today, stolen password collections are used as candidate passwords, which have become the foundation of password cracking.
- Not all attacks rely on technology; in fact, some cyberattacks use little if any technology to achieve their goals. Social engineering is a means of using trickery to cause the victim to act in the attacker's favor. Social engineering relies on an attacker's clever manipulation of human nature to persuade the victim to provide information or take action. Social engineering frequently uses impersonation or masquerading as a real or fictitious character and then playing out the role of that person on a victim. One of the most common attacks

based on social engineering is phishing. Phishing is sending an email or displaying a web announcement that falsely claims to be from a legitimate enterprise in an attempt to trick the user into surrendering private information or taking action. Typo squatting involves an attacker registering lookalike website domains so that users who misspell a web address will land on one of these domains. A hoax is a false warning, and attackers often send these out to trick users into taking risky actions.

- Social networking is the use of Internet-based social media platforms that allow users to stay connected with friends, family, or peers. The very nature of social networking (interacting electronically with a massive base of users whose true identities are hidden) carries a significant risk of becoming a target for attackers. Identity theft involves stealing another person's personal information, such as a Social Security number, and then using the information to impersonate the victim, often for financial gain.
- While the number of successful attacks may seem overwhelming to the point that no user can defend herself, that is not the case. All users can take basic steps for their protection, and these actions do not require an impregnable defense. The starting point is to first create a defensive stance that begins with this basic goal: the key to protecting my digital life is to make it too difficult for an attacker to upend my safety, financial security, and privacy.
- There are several defenses against password attacks. Strong passwords should be created for each separate account. Instead of relying on human memory for passwords, security experts today universally recommend that technology be used instead to create, store, and retrieve passwords. The technology used for securing passwords is called a password manager. In addition, two-factor authentication (2FA), using a special email account for receiving email reset links, creating an anonymous username, and using fictitious security answers can all enhance password security.
- Users must always be aware and be initially suspicious of virtually any email or electronic correspondence because it could be a social engineering attack. Several social engineering defenses can be used to prevent these attacks from being successful. Avoiding identity theft involves two basic steps. The first step is to deter thieves by safeguarding information. The second step is to monitor financial statements and accounts. Several general defenses can be used for any social networking site. First, users should be cautious about what information is posted on social-networking sites. Second, users should be cautious regarding who can view their information. Finally, because available security settings in social networking sites are often updated frequently by the site with little warning, users should pay close attention to information about new or updated security settings.

Key Terms

authentication
Fair and Accurate Credit Transactions Act (FACTA) of 2003
hoax
impersonation
online brute force attack
password
password crackers
password managers
password spraying
phishing
random password generator
social engineering
social networking
two-factor authentication (2FA)
typo squatting
weak password

Review Questions

1. What is the process of providing proof that a user is "genuine" and not an impostor?
  a. Registration
  b. Genuinization
  c. Validation
  d. Authentication

2. Which of the following is NOT an authentication credential?
  a. Something you exhibit
  b. Something you have
  c. Something you are
  d. Somewhere you have once lived

3. Which of the following is the result of scrambling a password through a one-way hash algorithm?
- **a.** MRD
- **b.** Quantium
- **c.** SLAM
- **d.** Message digest

4. What is relying on deceiving someone to obtain secure information?
- **a.** Sleight attack
- **b.** Social engineering
- **c.** Magic attack
- **d.** Brute force attack

5. What is the goal of a phishing attack?
- **a.** To capture keystrokes
- **b.** To send a fraudulent email to a user
- **c.** To duplicate a legitimate service
- **d.** To trick a user into surrendering personal information

6. Which of the following is NOT performed by an identity thief?
- **a.** Send malware into a bank's online accounting system.
- **b.** Open a bank account in the person's name and write bad checks on that account.
- **c.** Produce counterfeit checks or debit cards and then remove all money from the bank account.
- **d.** File for bankruptcy under the person's name to avoid paying debts they have incurred or to avoid eviction.

7. Which of the following is NOT a characteristic of a weak password?
- **a.** A common word
- **b.** A long password
- **c.** A predictable sequence of characters
- **d.** Text containing personal information

8. Which of the following is NOT a step to deter identity theft?
- **a.** Do not provide personal information either over the phone or through an email message.
- **b.** Keep personal information in a secure location.
- **c.** Carry a copy of a Social Security card in a wallet instead of the original.
- **d.** Shred financial documents that contain personal information.

9. Which of the following is NOT an attack on passwords?
- **a.** Password spraying attack
- **b.** Crack attack
- **c.** Using password collections
- **d.** Online brute force attack

10. How does a password cracker program work?
- **a.** Password crackers perform comparisons between unknown digests and a user's known password digest.
- **b.** Password crackers perform a one-way comparison between brute force and password sprayers.
- **c.** Password crackers perform comparisons between known digests and a user's unknown password digest.
- **d.** Password crackers perform algorithmic validations between known digests and unknown digests.

11. What type of attack redirects a user who enters americanbank.net into a web browser instead of the correct americanbank.com?
- **a.** Site redirection naming attack (SRNA)
- **b.** URL targeting
- **c.** Typo squatting
- **d.** Jack attacking

12. Which of the following attacks is rarely used today because of its low rate of success?
- **a.** Paint spray attack
- **b.** Phishing attack
- **c.** Cracker attack
- **d.** Online brute force attack

13. How can an attacker use a hoax?
- **a.** A hoax can trick a user into creating a short password.
- **b.** By sending out a hoax, an attacker can convince a user to read his email more often.
- **c.** A hoax could convince a user that malware is circulating and that he should change his security settings.
- **d.** A user who receives multiple hoaxes could contact his supervisor for help.

14. Why are long passwords stronger than short passwords?
- **a.** Long passwords are confusing to attackers who cannot read them.
- **b.** Long passwords require attackers to make many more attempts to uncover the password.
- **c.** Long passwords always use letters, numbers, and special characters so they are more puzzling to attackers.
- **d.** Short passwords take up less storage space, which makes them easier to break.

15. Which of the following principles is NOT used in a social engineering attack?
- **a.** Intimidation
- **b.** Consensus
- **c.** Unfamiliarity
- **d.** Authority

16. Which of the following is NOT a technique used in a social engineering attack to gain a user's trust?
- **a.** Project insecurity
- **b.** Use evasion and diversion
- **c.** Use humor
- **d.** Provide a reason

17. Which of the following is true about phishing?
- **a.** The rates of phishing attacks are on the decline.
- **b.** Phishing is sending an email or displaying a web announcement that falsely claims to be from a legitimate enterprise.
- **c.** Phishing can only be performed through text messaging.
- **d.** Phishing is not considered to be a social engineering attack.

18. Why is it difficult to stop typo squatting attacks?
- **a.** The large number of gTLDs makes it hard for organizations to register sites that are a variation of their registered domain name.
- **b.** The accurate spelling of users has continued to decline.
- **c.** There is no known defense against these attacks.
- **d.** A change in Internet protocols had made these attacks more successful.

19. Which of the following is NOT a risk associated with using a social networking account?
- **a.** Accepting friends may have unforeseen consequences.
- **b.** Social networking security is confusing.
- **c.** Users may not be trusting of others.
- **d.** Personal data can be used maliciously.

20. Which of the following is NOT considered a good password defense?
- **a.** Turn off 2FA.
- **b.** Use a special email account for email reset links.
- **c.** Use an anonymous username.
- **d.** Create fictitious security answers.

Hands-On Projects

Caution !

If you are concerned about installing any of the software in these projects on your regular computer, you can instead use the Windows Sandbox created in Module 1 Hands-On Projects. Software installed within the virtual machine will not impact the host computer.

Project 2-1: Using an Online Password Cracker

In this project, you create a digest on a password and then crack it to demonstrate the speed of cracking weak passwords.

1. The first step is to use a general-purpose hash algorithm to create a password hash. Use your web browser to go to **www.fileformat.info/tool/hash.htm**. (The location of the content on the Internet may change without warning; if you are no longer able to access the program through this URL, use a search engine and search for "Fileformat.info.")
2. Under **String hash**, enter the simple password **apple123** in the **Text:** box.
3. Click **Hash**.
4. Scroll down the page and copy the MD5 hash of this password to your Clipboard by selecting the text, right-clicking it, and choosing **Copy**.
5. Open a new tab on your web browser.
6. Go to **https://crackstation.net/**.
7. Paste the MD5 hash of *apple123* into the text box below **Enter up to 20 non-salted hashes, one per line**.
8. In the reCAPTCHA box, click the **I'm not a robot** check box.
9. Click **Crack Hashes**.
10. How long did it take to crack this hash?
11. Click the browser tab to return to FileFormat.Info.
12. Under **String hash**, enter the longer password **applesauce1234** in the **Text:** box.

Continued

13. Click **Hash**.
14. Scroll down the page and copy the MD5 hash of this password to your Clipboard.
15. Click the browser tab to return to the CrackStation site.
16. Paste the MD5 hash of *applesauce1234* into the text box below **Enter up to 20 non-salted hashes, one per line**.
17. In the reCAPTCHA box, click the **I'm not a robot** check box.
18. Click **Crack Hashes**.
19. How long did it take this online rainbow table to crack this stronger password hash?
20. Click the browser tab to return to FileFormat.Info and experiment by entering new passwords, computing their hash, and testing them on the CrackStation site. If you are bold, enter a string hash that is similar to a real password that you use.
21. What does this tell you about the speed of password-cracking tools? What does it tell you about how easy it is for attackers to crack weak passwords?
22. Close all windows.

Project 2-2: Using a Web-Based Password Manager

The drawback to using strong passwords is that they can be very difficult to remember, particularly when a unique password is used for each account that a user has. As another option, password manager programs allow users to create, store, and retrieve account information. One example of a web-based password storage program is Bitwarden, which is an open-source product. In this project, you create an account and use Bitwarden.

1. Use your web browser to go to **https://bitwarden.com/**. (The location of the content on the Internet may change without warning; if you are no longer able to access the program through this URL, use a search engine and search for "bitwarden.")
2. Click **View Plans & Pricing**, and then click the **Personal** tab.
3. Under **Basic Free Account**, click **Create Free Account**.
4. Enter the requested information and then click **Submit**.
5. Close Bitwarden.
6. Now go to your Bitwarden login page at **https://vault.bitwarden.com/#/vault**.

Caution !

The page from which you access your Bitwarden account is not the same as the page used for creating the account.

7. Enter your email address as your username and your master password.
8. Click **Log in**.
9. To add an account, click **Add item**.
10. Under **What type of item is this?**, select **Login**.
11. Enter the requested information as it applies to this account. Under **Name**, enter the name of this account. Under **Password**, note that a random password generator is available. Leave the **Authenticator Key (TOTP)** blank. Under **Match Detection**, select **Default match detection**.
12. The **Notes** section can be used for other information such as your random answers to the security questions for this site.
13. Click **Save**.
14. Now use Bitwarden to access this account. Hover over the account name in **My Vault** to display the gear icon. (This is different from the gear icon next to **Add Item**.)
15. From the drop-down menu, click **Launch**. You are now taken to this site.

16. Return to Bitwarden.
17. Hover over the account name in **My Vault** to display the gear icon and click **Copy Username**.
18. Return to your account and paste this username into the field.
19. Hover over the account name in **My Vault** to display the gear icon and click **Copy Password**.
20. Return to your account and paste the password into the field.
21. Log out of this account.
22. Now return to Bitwarden and enter a second account that you frequently use, and then practice going to that account with the information stored in Bitwarden.
23. Log out of this account.
24. How easy is Bitwarden to use? Can you see how it can provide stronger security than memorizing weak passwords?
25. Log out of Bitwarden.
26. Close all windows.

Project 2-3: Using a Password Manager Application

The drawback to using a web-based password manager is that its entire security depends on the strength of the website. For this reason, some security professionals prefer to use an application that is downloaded and stored on their computer instead. One example of a password manager application is KeePass Password Safe, which is an open-source product. In this project, you download and install KeePass.

1. Use your web browser to go to **https://keepass.info/**. (The location of the content on the Internet may change without warning; if you are no longer able to access the program through this URL, use a search engine and search for "keepass.")
2. Under **Getting KeePass**, click **Downloads**. Locate the most recent version of Installer for Windows and click it to download the application. Save this file in a location such as your desktop, a folder designated by your instructor, or your portable USB flash drive. When the file finishes downloading, install the program. Accept the installation defaults.
3. Launch KeePass to display the opening screen.
4. Click **File** and then click **New** to start a password database. Enter a strong master password for the database to protect all the passwords in it. When prompted, enter the password again to confirm it.
5. Create a group by clicking **Group** and clicking **Add Group**. Enter **Web Sites** and click **OK**.
6. Select the **Web Sites** group, click **Entry**, and then click **Add Entry**.
7. Enter a title for a website you use that requires a password under **Title**.
8. For **User name**, enter the username that you use to log in to this account.
9. Delete the **Password** and **Repeat** entries, enter the password that you use for this account, and then confirm it.
10. Enter the URL for this account in the **URL** box.
11. Click **OK**.
12. Click **File** and then click **Save**. Enter your last name as the file name and then click **Save**.
13. Exit KeePass.
14. Launch the application.
15. Enter your master password to open your password file.
16. If necessary, click the group to locate the account you just entered; it will be displayed in the right pane.
17. Click the Title of the site you just created.
18. Click the **Open URL(s)** icon in the toolbar to go to this site.
19. Click in the username field of this site that asks for your username.
20. Return to KeePass.
21. Click the **Perform Auto-Type** icon in the toolbar.

Continued

Note 20

The Perform Autotype feature will automatically enter your username and password into these fields in your online account. However, depending on how the website has been designed, they may not automatically populate these fields. As an alternative, you can drag and drop the username and password from KeePass into the account fields.

22. Because you can drag and drop your account information from KeePass, you do not have to memorize any account passwords and can instead create strong passwords for each account. Is this an application that would help users create and use strong passwords? What are the strengths of this program? What are the weaknesses? Would you use KeePass?
23. Close all windows.

Project 2-4: Viewing Your Annual Credit Report

Security experts recommend that one means of reducing personal risk for consumers is to receive a copy of their credit report at least once per year and check its accuracy to protect their identity. In this project, you access your free credit report online.

1. Use your web browser to go to **www.annualcreditreport.com**. Although you could send a request individually to one of the three credit agencies, this website acts as a central source for ordering free credit reports.
2. Click **Request your free credit reports**.
3. Read through the three steps and click **Request your credit reports**.
4. Enter the requested information, click **Continue**, and then click **Next**.
5. Click **TransUnion**. Click **Next**.
6. After the brief processing completes, click **Continue**.
7. You may then be asked for personal information about your transaction history to verify your identity. Answer the requested questions and click **Next**.
8. Follow the instructions to print your report.
9. Review it carefully, particularly the sections of "Potentially negative items" and "Requests for your credit history." If you see anything that might be incorrect, follow the instructions on that website to enter a dispute.
10. Follow the instructions to exit from the website.
11. Close all windows.

Project 2-5: Online Phishing Training

In this project, you will use an online phishing training tool. Also, note the user awareness training features in this simulation as you proceed.

1. Use your web browser to go to **https://public.cyber.mil/training/phishing-awareness/**. (If you are no longer able to access the program through this URL, use a search engine and search for "phishing awareness.")
2. Click **Launch Training**.
3. If necessary, adjust your web browser settings, and then click **Start/Continue Phishing and Social Engineering: Virtual Communication Awareness**.
4. Watch the brief video on accessibility features. Click the right arrow button.
5. Read the information. Click either the URL or **Continue** depending upon your needs.
6. Listen to the video message about your choice. Is this a good learning technique? Why? Click the right arrow button.

7. Continue through the phishing training. Slides 16–18 ask you for answers to questions about what you have learned.
8. How effective was this training? What did you learn? Would you recommend this to others to learn about phishing?
9. Close all windows.

Case Projects

Case Project 2-1: Testing Password Strength

How strong are your passwords? Various online tools can provide information on password strength, but not all feedback is the same. First, assign the numbers 1–3 to three passwords that are very similar (but not identical) to passwords you are currently using, and write down the number (not the password) on a piece of paper. Then, enter those passwords into these three online password testing services:

- How Secure Is My Password (www.security.org/how-secure-is-my-password/)
- Password Checker Online (http://password-checker.online-domain-tools.com/)
- The Password Meter (www.passwordmeter.com/)

Next to each number, record the strength of that password as indicated by these three online tools. Then use each online password tester to modify the password by adding more random numbers or letters to increase its strength. How secure are your passwords? Would any of these tools encourage someone to create a stronger password? Which provided the best information? Create a one-paragraph summary of your findings.

Case Project 2-2: Password Requirements

Visit the website Passwords Requirements Shaming (password-shaming.tumblr.com), which is a list of password requirements for different websites that are considered weak. Read through several of the submissions. Select three that you consider the most egregious. Why are they the worst? Next, indicate what you would suggest for making the requirement stronger, but a requirement that most users could meet. Write a one-paragraph summary.

Case Project 2-3: Protecting Yourself from Identity Theft

Use the Internet to identify at least three different sources of information about preventing identity theft. Then create a table that lists at least 10 steps that users can take to protect themselves and a detailed description of each step. Finally, share your document with at least two other individuals. Ask them to rank each step on a scale from 1–5 indicating how likely they would be to follow it. Write a one-paragraph summary of your findings.

Case Project 2-4: Social Networking Assessment

Assess the security of one of your social networking sites. Begin by looking at the settings that you have on this site. Then explore the different security configurations that you could add to your site to make it more secure. What about the different users who have access to your postings? Is this something that you should pare down? Write a one-paragraph paper on what you found and what you could do to increase the security of your social networking site.

Continued

Case Project 2-5: Your Defensive Stance

Evaluate your defensive stance. In particular, answer these three questions:

- What do I need to protect? (That is, what in my digital life can give away critical information tied to my finances, privacy, and safety?)
- How likely is it that it needs protection? (That is, what is your current personal level of exposure to threats?)
- Will the effort be worth it to protect my digital life? (That is, do you want to spend the energy to protect yourself?)

Write a one-paragraph analysis of your defensive stance and how it could be improved.

References

1. Croce, Brian, "Cybersecurity 'patchwork' leaving industry vulnerable," *Pensions & Investments*, Feb. 4, 2019, accessed Apr. 16, 2022, https://www.pionline.com/article/20190204/PRINT/190209947/cybersecurity-patchwork-leaving-industry-vulnerable.
2. Le Bras, Tom, "Online overload – it's worse than you think," *Dashlane Blog*, accessed Apr. 26, 2022, https://blog.dashlane.com/infographic-online-overload-its-worse-than-you-thought/.
3. Ruef, Marc, "Password leak analysis: extensive analysis of passwords," SCIP, Apr. 15, 2021, retrieved Apr. 17, 2022, https://www.scip.ch/en/?labs.20210415.
4. Schneier, Bruce, *Secrets and lies: Digital Security in a Networked World* (New York: Wiley Computer Publishing), 2004.
5. "Phishing attacks hit all-time high in December 2021," *Help Net Security*, Mar. 3, 2022, accessed Apr. 18, 2022, https://www.helpnetsecurity.com/2022/03/03/phishing-attacks-december-2021/#:~:text=APWG%20saw%20316%2C747%20phishing%20attacks,has%20tripled%20from%20early%202020.
6. Lua, Alfred, "20 Top social media sites to consider for your brand in 2022," *Buffer Marketing Library*, accessed Apr. 18, 2022, https://buffer.com/library/social-media-sites/.

Module 3

Computer Security

After completing this module, you should be able to do the following:

1. Define malware.
2. Identify the different types of malware attacks.
3. Explain how managing patches and running antimalware software can provide a defense.
4. Explain what a firewall does.
5. Describe how to stop ransomware.

Cybersecurity Headlines

In December 2019, residents of Wuhan City, Hubei Province, China began experiencing respiratory illnesses. Technically, it was designated as severe acute respiratory syndrome coronavirus 2 (SARS-CoV-2) but more commonly it was called COVID-19 (*co*rona*vi*rus *d*isease 20*19*). The outbreak was initially reported to the World Health Organization (WHO) on the last day of 2019. By the end of January 2020, the WHO had declared COVID-19 a global health emergency. Six weeks later (March 11, 2020), the WHO declared COVID-19 a global pandemic, its first such designation since declaring H1N1 influenza a pandemic in 2009. In just over two years since that time, there have been over 500 million global confirmed cases accounting for over 6 million deaths, with almost 1 million of those deaths occurring in the United States.

How was COVID-19 able to grow so quickly to become a global pandemic? Part of the answer lies in the fact that COVID-19 was new. Although coronaviruses are a large family of viruses that can cause illnesses ranging from the common cold to more severe diseases, a "novel" coronavirus (nCoV) like COVID-19 is a new strain that has not been previously identified.

The newness of COVID-19 was impactful in two ways. First, COVID-19 was new to epidemiologists (from three Greek words epi ("upon"), demos ("people"), and logos ("study")) who study the distribution and causes of health-related states and events. These scientists had not seen this coronavirus before and thus were not prepared for it. Second, it was new to our human bodies. Because this outbreak was new, our bodies had no ready defenses against it. In short, because COVID-19 was new, it rapidly spread around the world with no known defenses to stop it.

In a remarkably similar way to COVID-19, the most frightening cybersecurity malware that attacks computers takes advantage of vulnerabilities that are completely new and entirely unknown. Cybersecurity attacks that use previously unknown vulnerabilities are called "zero-day attacks," because they give zero days of advanced warning. And like COVID-19, zero-day attacks result in large numbers of end-users with infected computers and security professionals scrambling to create a defense.

How widespread are zero-day attacks? Google's Project Zero, which tracks these vulnerabilities, has reported a sharp increase. For example, in 2020, only 25 zero-day vulnerabilities were used by attackers. One

Continued

year later, that number more than doubled to 58. In 2014, the year records were first kept by Project Zero, there were only 14 vulnerabilities. To date, attackers have taken advantage of 215 previously unknown vulnerabilities to launch attacks.

Several elements make zero-day attacks especially dangerous. First, they are eagerly sought after and highly prized. Because they are completely unknown by the cybersecurity community, thousands or hundreds of thousands of successful attacks may occur before the vulnerability is uncovered and a fix can be distributed. Second, it is suspected that zero-day vulnerabilities are being "stockpiled" by attackers and are being "pulled out" to be used when the time is right. Finally, it is not only attackers who are using these vulnerabilities, but governments as well are constantly looking for zero-day vulnerabilities to use against other nations or even their own citizens.

The danger of zero-day attacks, like unknown viruses, will likely continue to be with us well into the future.

Protecting your personal computing device—be it a desktop, laptop, or tablet—is a challenge, even for the most advanced computer users. This is because many different types of attacks are launched against these devices, and attackers are constantly modifying these attacks as well as creating new ones daily.

Just as a house must be protected against different types of threats—burglary, arson, vandalism, hurricanes, mold, and termites—so too must a computer be protected from a variety of different attacks. And just as protecting against termites is much different from protecting against burglary, several defenses must be in place for a computer to remain safe.

In this module, you will learn about computer security. You will start by looking at the types of computer attacks that occur today, and then learn what defenses must be in place to keep computers and the information stored on them secure.

Malware Attacks

Most successful attacks on computing devices are a result of malware. **Malware** (*mal*icious soft*ware*) is software that enters a computer system without the user's knowledge or consent and then performs an unwanted and harmful action.

Note 1

Malware is most often used as a general term that refers to a wide variety of damaging software programs.

The number of instances of malware is staggering. According to one report, the number of new malware instances every month exceeds 20 million, and the total malware instances in existence are approaching 900 million.[1] Although security protections continue to improve, malware likewise is continually evolving to avoid detection and successfully attack computers.

There is no standard established for the classification of the different types of malware as there is for infectious diseases like coronaviruses. This often results in user confusion over malware, since it is difficult to know the similarities and differences between different types of malware.

One attempt at classifying the diverse types of malware can be to examine the *primary action that the malware performs* and then group those together that have similar actions. When classifying malware in this fashion, these malware actions used for groupings can be kidnap, eavesdrop, masquerade, serve as a launchpad, and sidestep.

Note 2

Some malware performs more than one of these actions. However, in terms of classification, the *primary* action of the malware is used here.

Kidnap

Kidnapping is a crime that involves capturing a person and then holding them captive until a ransom is paid for their release. Paying a ransom for a kidnapped victim has a long history. The Roman general and statesman Julius Caesar (100 BC–44 BC) was a kidnapping victim. Caesar was captured near the island Pharmacusa by pirates. His captors demanded 20 talents as his ransom, but Caesar laughed at them for not knowing who he was and said the ransom should be 50 talents! Once the ransom was paid and Caesar was set free, he immediately set out to find his captors. After catching up with them he arrested them and declared the ransom as his own "booty" so he could keep it for himself.

Note 3

Although sometimes used interchangeably, there is a difference between kidnapping, abduction, and holding hostage. Kidnapping is seizing and imprisoning a person against their will to receive a ransom for their release. Abducting occurs when a person has been taken away from their original location by persuasion, fraud, force, or even violence but there is no intent to exchange the person for money. Holding hostage involves a person held by a captor to force the government to meet certain conditions, such as the release of prisoners.

Children are often the victims of kidnapping (hence, the term *kid*napping) since they are unable to defend themselves. Charles Lindbergh's 20-month-old son was kidnapped in 1932 from the nursery on the second floor of the Lindbergh home in New Jersey and a total of 13 ransom notes were received, demanding a ransom that started at $70,000 but was later reduced to $50,000. The body of the child was later found. In 1968, Barbara Jane Mackle, the daughter of a wealthy Florida land developer, was kidnapped and buried in a fiberglass box while the kidnappers demanded half a million dollars, which was eventually paid. She was rescued after spending 83 hours underground. Recently, several African nations are experiencing mass kidnappings of hundreds of children from schools and hospitals.

Note 4

It is estimated that in one year, over 3,000 people have been kidnapped across Nigeria.

Similarly, attackers today perform a "kidnapping" on a user's computer and hold it "hostage" until a ransom is paid. **Ransomware** is the malicious software designed to extort money from victims in exchange for their computer being restored to its normal working state.

There are two general types of ransomware. The first is known as blocker ransomware while the other is called cryptomalware.

Blocker Ransomware

The earliest form of ransomware is **blocker ransomware**. Blocker ransomware prevents users from using their computers in a normal fashion. This occurs by the ransomware infecting the computer and then manipulating its operating system (Microsoft Windows, Apple macOS, or Linux) to block all normal access to the device.

Note 5

Ransomware first became widespread around 2010. It is the fastest-growing malware in existence today.

Typically, once blocker ransomware infects a computer, then a message on the user's screen appears pretending to be from a reputable third party. This message usually provides a "valid" reason for blocking the user's access to the computer. One common example is ransomware that purports to come from a law enforcement agency. The message, using official-looking imagery, states that the user has performed an illegal action such as downloading pirated software and must now immediately pay a fine online by entering a credit card number. Once the fine is paid, the message says, the computer will be restored to its normal function. Figure 3-1 shows a blocker ransomware message.

Figure 3-1 Blocker ransomware message

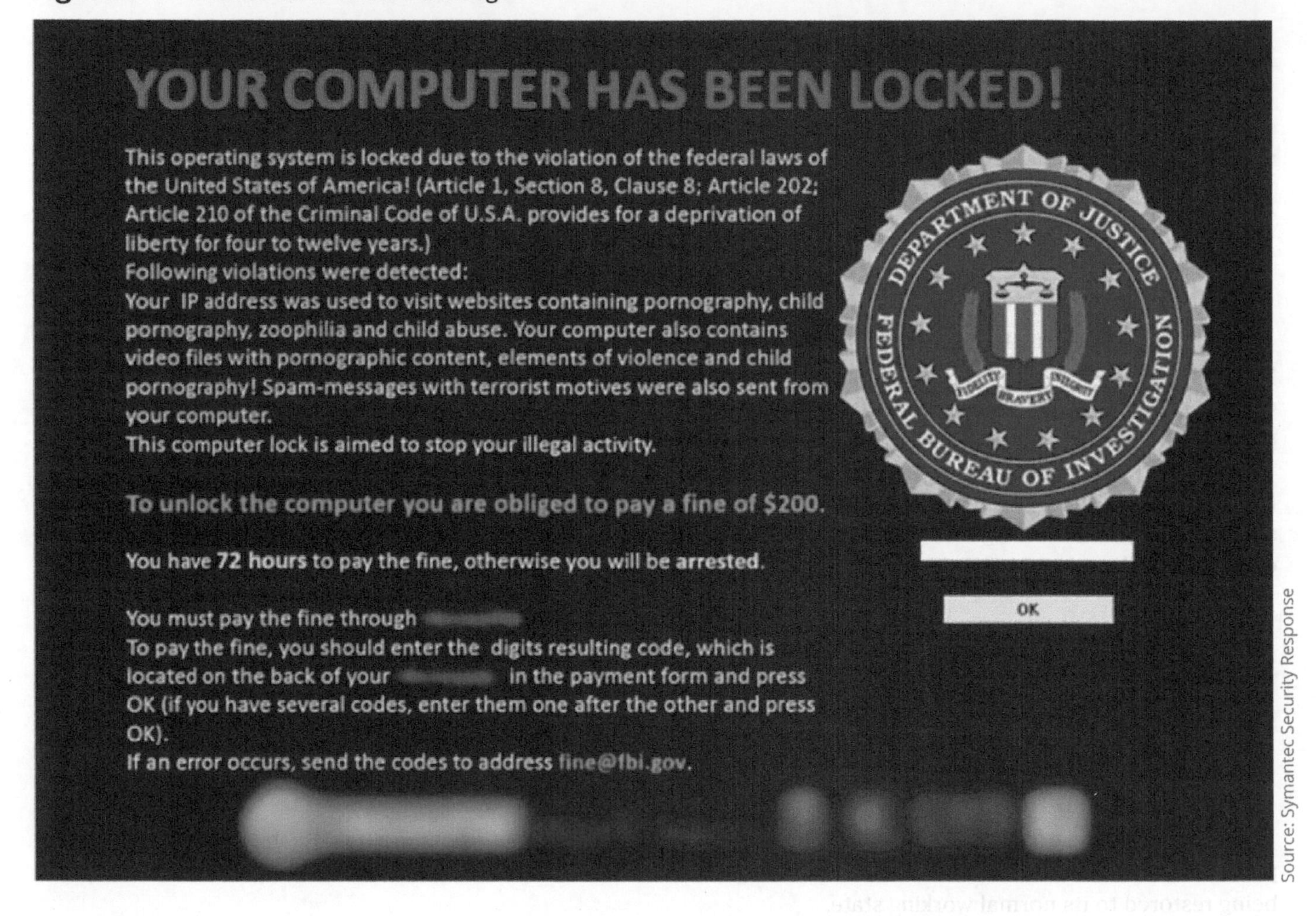

Source: Symantec Security Response

Caution !

Ransomware embeds itself onto the computer so that it cannot be bypassed. Even rebooting over and over has no impact.

Another variation of blocker ransomware pretends to come from a reputable software vendor. It displays a fictitious warning that a software license has expired or the computer has a problem such as imminent hard drive failure or—in a touch of irony—a malware infection. This ransomware variation tells users that they must immediately renew their license or purchase additional software online to fix a non-existent problem. The ransomware example in Figure 3-2 uses color schemes and icons like those found on legitimate software.

Figure 3-2 Blocker ransomware variation computer infection

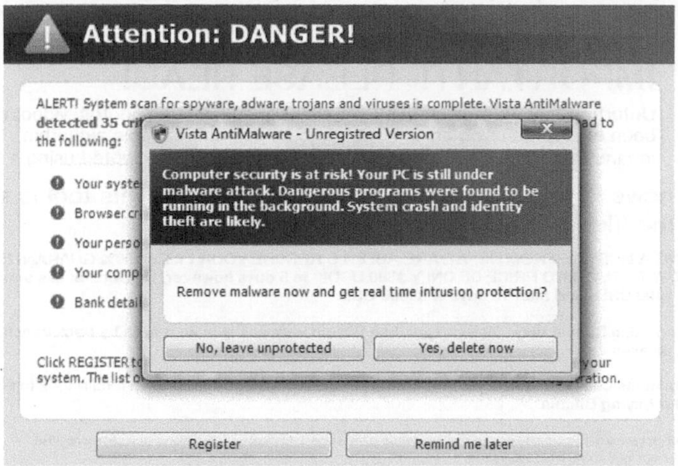

Source: Microsoft Corporation

As ransomware has become more widespread, attackers today often drop the pretense that the ransomware is from a reputable third party. Instead, they simply block the user's computer and demanded a fee for its release.

Note 6

Ransomware attackers have determined what they consider the optimal price point for payment by users to unblock a computer: the amount must be small enough that most victims will begrudgingly pay to have their systems unblocked, but large enough that when thousands of victims pay up, the attackers can garner a handsome sum. For individuals, the ransom is usually around $500. However, for enterprises, the price can be tens or even hundreds of millions of dollars. Some ransomware attackers have even set up help desks that victims can call to receive assistance in paying the ransom!

Ransomware continues to be a serious threat. In 2021, about 37 percent of all global organizations were the victim of some form of ransomware attack, which is a 62 percent increase over the previous year. There were ransomware incidents against 14 of the 16 U.S. critical infrastructure sectors. Since 2020, more than 130 different ransomware strains have been detected.[2]

Note 7

The top four ransomware targets are education, retail, business and professional services, and government. Threat actors often target state and local governments because they typically have weaker security. To date, over 469 of these governments have been the victims of successful attacks.[3]

Cryptomalware

In recent years, a more malicious form of ransomware has arisen. Instead of just blocking the user from accessing the computer, this ransomware encrypts some or all the files on the device so that they cannot be opened. (Encrypting only some files helps the malware to evade detection.) This is called **cryptomalware**.

In a cryptomalware attack, after the files have been encrypted, a message appears telling the victim that his files are now encrypted, and a fee must be paid to receive a key to unlock them. In addition, the message often contains a warning about the increased urgency for payment by claiming that the cost for the key to unlock the cryptomalware increases every few hours or days. On some occasions, the threat actors claim that an ever-increasing number of the encrypted user files will be deleted until the ransom is paid. If the ransom is not paid by a specific deadline, then the key to unlock the files can never be purchased. Figure 3-3 shows a cryptomalware message.

Figure 3-3 Cryptomalware message

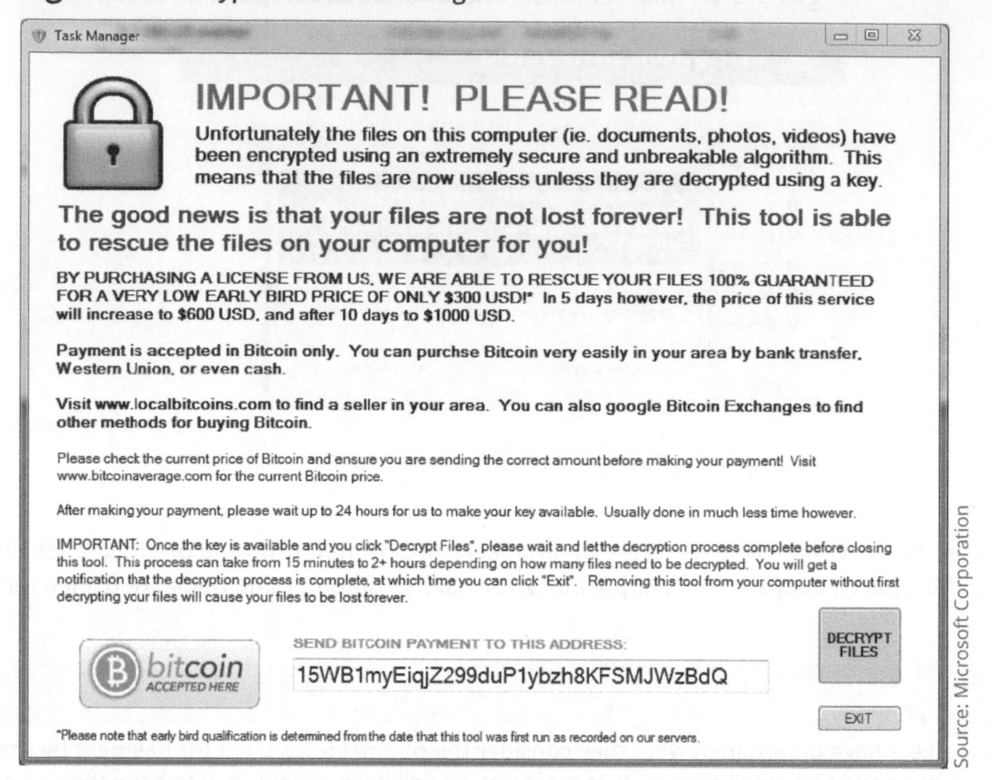

Source: Microsoft Corporation

Note 8

With early cryptomalware attacks, threat actors only delivered the decryption key fewer than half the times that a ransom was paid. However, this resulted in many victims not paying the ransom since word circulated that the risk was high of not getting the key. Threat actors have since learned that they have more to gain in the long run by making the key available after a ransom is paid. Today when victims pay the ransom, a decryption tool is delivered 99 percent of the time. However, the key only works about 96 percent of the time. This is because specific variants of ransomware tend to corrupt data when it is encrypted.[4]

In addition to encrypting files on the user's local hard drive, new variants of cryptomalware encrypt all files on *any* network or device connected to that computer. This includes secondary hard disk drives, USB hard drives, network-attached storage devices, network servers, and even cloud-based data repositories.

Note 9

Originally, the FBI did not support paying a ransom in any circumstances. It said, "The FBI does not advocate paying a ransom, in part because it does not guarantee an organization will regain access to its data. . . Paying ransoms emboldens criminals to target other organizations and provides an alluring and lucrative enterprise to other criminals." However, the FBI later seemingly softened its stance by adding, "However, the FBI understands that when businesses are faced with an inability to function, executives will evaluate all options to protect their shareholders, employees, and customers."[5]

Eavesdrop

Another category of malware eavesdrops or secretly listens to ("spies on") its victims. The two common types of eavesdropping malware are keyloggers and spyware.

Keylogger

A **keylogger** silently captures and stores each keystroke that a user types on the computer's keyboard. The threat actor can then search the captured text for any useful information such as passwords, credit card numbers, or personal information. A keylogger can be a software program or a small hardware device.

Software keyloggers are programs installed on the computer that silently capture sensitive information. However, software keyloggers, which conceal themselves so that the user cannot detect them, go far beyond capturing a user's keystrokes. These programs can also make screen captures of everything on the user's screen and silently turn on the computer's web camera to record images of the user. A software keylogger is illustrated in Figure 3-4.

Figure 3-4 Software keylogger

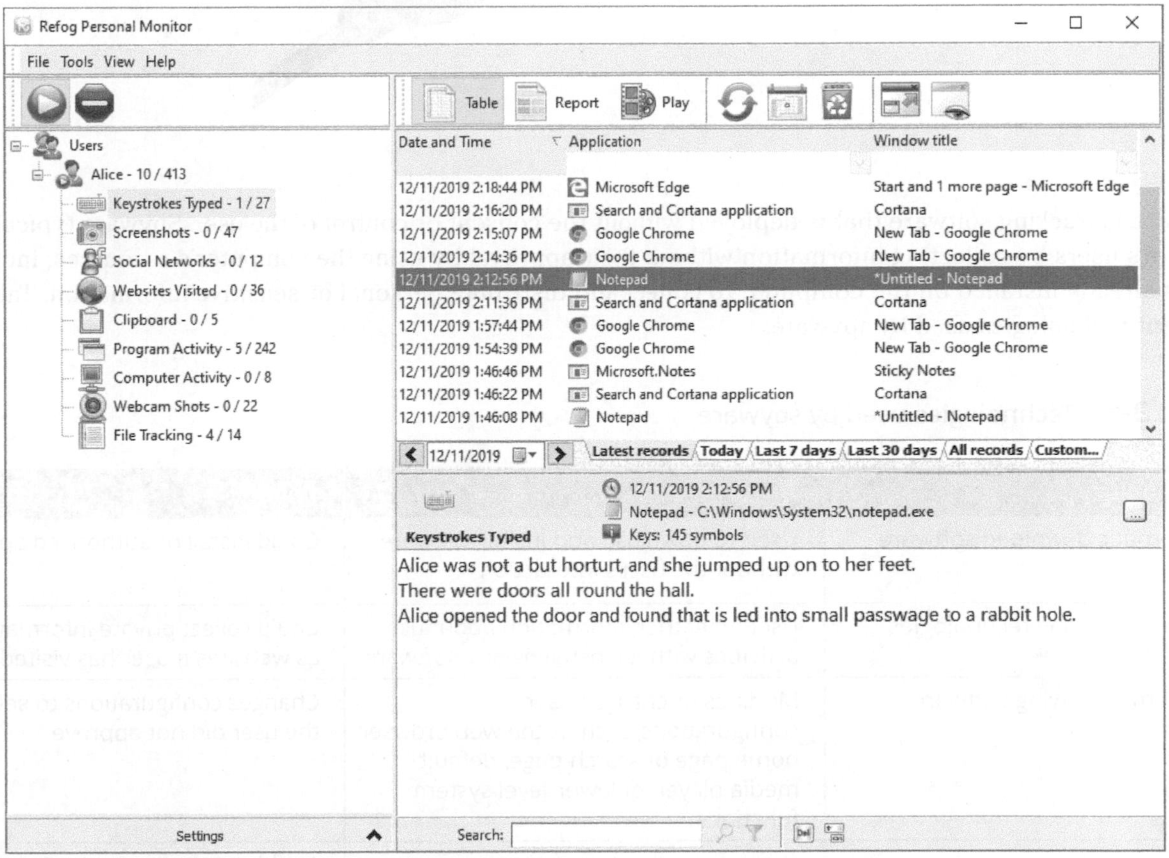

Source: Refog

Note 10

An advantage of software keyloggers is that they do not require physical access to the user's computer. This malware can be installed remotely and then routinely send captured information back to the attacker through the victim's Internet connection.

For computers that are in a public location such as a library or computer lab but are "locked down" so that no software can be installed, a hardware keylogger can be used instead. These keyloggers are hardware devices inserted between the computer keyboard connection and USB port, as shown in Figure 3-5. Because the device resembles an ordinary keyboard connection and the computer keyboard USB port is often on the back of the computer, a hardware keylogger can easily go undetected. In addition, the device is beyond the reach of the computer's antimalware scanning software and thus raises no alarms. A disadvantage of a hardware keylogger is that the threat actor must install and then later return to physically remove the device to access the information it has stored, each time being careful not to be detected.

Figure 3-5 Hardware keylogger

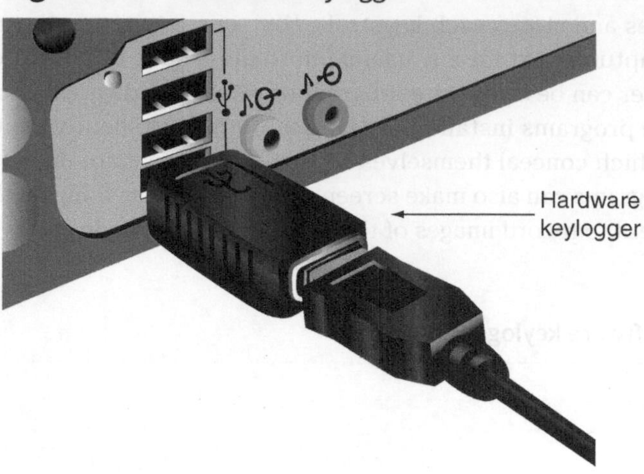

Spyware

Spyware is tracking software that is deployed without the consent or control of the user. Spyware typically secretly monitors users by collecting information without their approval by using the computer's resources, including programs already installed on the computer, to collect and distribute personal or sensitive information. Table 3-1 lists different technologies used by spyware.

Table 3-1 Technologies used by spyware

Technology	Description	Impact
Automatic download software	Used to download and install software without the user's interaction	Could install unauthorized applications
Passive tracking technologies	Used to gather information about user activities without installing any software	Could collect private information such as websites a user has visited
System modifying software	Modifies or changes user configurations, such as the web browser home page or search page, default media player, or lower-level system functions	Changes configurations to settings that the user did not approve
Tracking software	Used to monitor user behavior or gather information about the user, sometimes including personally identifiable or other sensitive information	Could collect personal information that can be shared widely or stolen, resulting in fraud or identity theft

Caution !

Not all spyware is necessarily malicious. For example, spyware monitoring tools can help parents keep track of the online activities of their children.

Masquerade

Some malware attempts to deceive the user and hide its true intentions by "masquerading" or pretending to be something else. Software in this category includes potentially unwanted programs (PUPs), Trojans, and Remote Access Trojans (RATs).

Potentially Unwanted Program (PUP)

A broad category of software that is often more annoying than malicious is called **potentially unwanted programs (PUPs)**. A PUP is software that the user does not want on their computer. PUPs are often installed along with other programs as a result of the user overlooking the default installation options on software downloads. As shown in Figure 3-6, multiple check boxes are default installation options so that when the user downloads the main program, these PUPs will also be downloaded and installed.

Figure 3-6 Default installation options

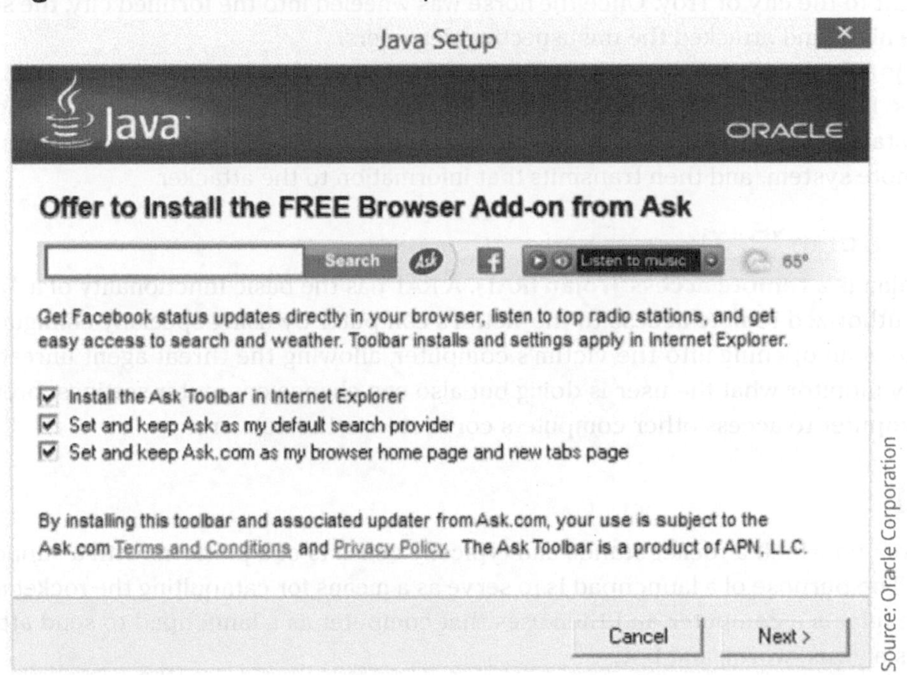

Source: Oracle Corporation

PUPs may include software that is preinstalled on a new computer or smartphone (called "bloatware") and cannot be easily removed (if at all). Many PUPs display advertising through pop-up windows that obstruct content, as shown in Figure 3-7.

Figure 3-7 Pop-up window

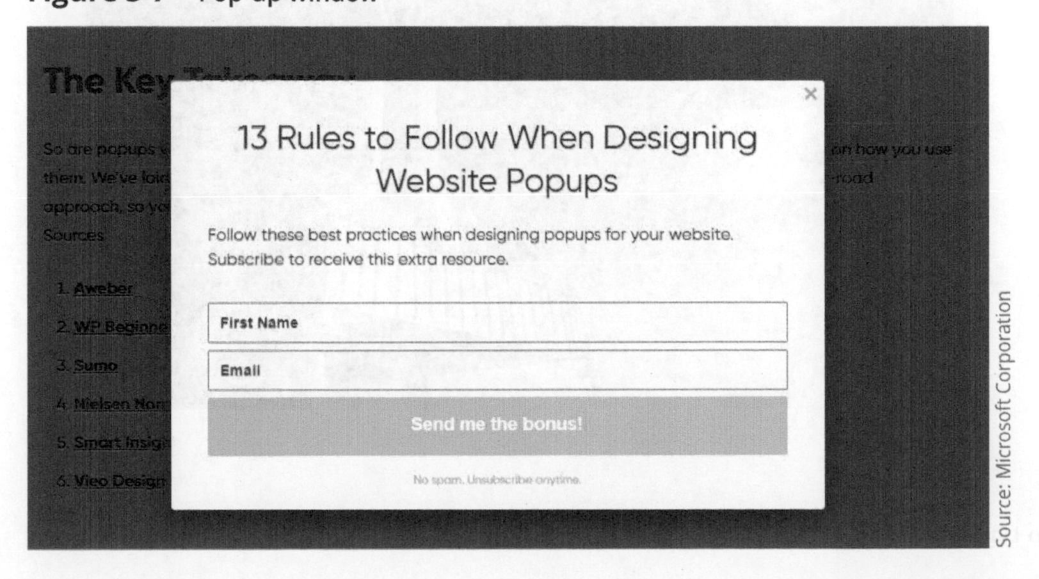

Source: Microsoft Corporation

Note 11

The term PUP was created by an Internet security company because marketing firms objected to having their products being called "spyware."

Trojan

According to ancient legend, the Greeks won the Trojan War by hiding soldiers in a large, hollow, wooden horse that was presented as a gift to the city of Troy. Once the horse was wheeled into the fortified city, the soldiers crept out of the horse during the night and attacked the unsuspecting defenders.

A computer **Trojan** is an executable program that masquerades as performing a benign activity but also does something malicious. For example, a user might download what is advertised as a calendar program, yet installing the calendar also installs malware that scans the system for credit card numbers and passwords, connects through the network to a remote system, and then transmits that information to the attacker.

Remote Access Trojan (RAT)

A special type of Trojan is a **remote access Trojan (RAT)**. A RAT has the basic functionality of a Trojan but also gives the threat agent unauthorized remote access to the victim's computer by using specially configured communication protocols. This creates an opening into the victim's computer, allowing the threat agent unrestricted access. The attacker can not only monitor what the user is doing but also can change computer settings, browse and copy files, and even use the computer to access other computers connected to the network.

Launchpad

A launchpad is an area on which a rocket stands and typically consists of a platform with a supporting structure, as shown in Figure 3-8. The purpose of a launchpad is to serve as a means for catapulting the rocket into space. Another category of malware infects a computer and then uses that computer as a launchpad to send attacks to other computers. This includes a virus, worm, and bot.

Figure 3-8 Rocket launchpad

Source: 3Dsculptor/Shutterstock.com

Virus

There are two types of viruses. These are a file-based virus and a fileless virus.

File-based Virus A biological virus is composed of tiny bits of genetic material enclosed by a protective shell. By themselves, viruses are lifeless and inert, waiting for a favorable environment in which to reproduce. When a virus encounters a host cell, the virus attaches itself to the outer wall of the cell, enters inside, travels to the cell's genome, merges with its genes, and then tricks the host's genome into making copies of itself.

Note 12

When the host cell is infected by a virus, the virus takes over the operation of that cell, converting it into a virtual factory to make more copies of it. The host cell will produce millions of identical copies of the original virus very rapidly. Biologists often say that viruses exist only to make more viruses.

A **file-based virus** is remarkably similar to a biological virus. It is malicious computer code attached to a file. Each time the infected program is launched or the data file is opened—either by the user or the computer's operating system—the virus first unloads a payload to perform a malicious action (corrupt or delete files, prevent programs from launching, steal data to be sent to another computer, cause a computer to crash repeatedly, or turn off the computer's security settings). Then the virus reproduces itself by inserting its code into another file, but only on the same computer. A virus can only replicate itself on the host computer on which it is located; it cannot automatically spread to another computer by itself. Instead, it must rely on the actions of users to spread to other computers.

Because viruses are attached to files, they can be spread when a user transfers those files to other devices. For example, a user might send an infected file as an email attachment or copy an infected file to a USB flash drive and give the drive to another user. Once the virus reaches a new computer, it begins to infect it. Thus, a virus must have two carriers: a file to which it attaches and a human to transport it to other computers.

Many file types can contain a virus. Table 3-2 lists some of the 50 different Microsoft Windows file types that can be infected with a virus.

Table 3-2 Windows file types that can be infected

File extension	Description
.docx or .xlsx	Microsoft Office user documents
.exe	Executable program file
.msi	Microsoft installer file
.msp	Windows installer patch file
.scr	Windows screen saver
.cpl	Windows Control Panel file
.msc	Microsoft Management Console file
.wsf	Windows script file
.ps1	Windows PowerShell script

Note 13

One of the first viruses found on a microcomputer was written for the Apple II in 1982. Rich Skrenta, a ninth-grade student in Pittsburgh, wrote "Elk Cloner," which displayed his poem on the screen after every 50th use of the infected floppy disk. Unfortunately, the virus leaked out and found its way onto the computer used by Skrenta's math teacher. In 1984, the mathematician Dr. Frederick Cohen introduced the term *virus* based on a recommendation from his advisor, who came up with the name from reading science fiction novels.

Fileless Virus A **fileless virus**, on the other hand, does not attach itself to a file. Instead, fileless viruses take advantage of native services and processes that are part of the operating system to avoid detection and carry out their attacks. The native services used in a fileless virus are called *living-off-the-land binaries (LOLBins).* For a computer running Microsoft Windows, some commonly exploited LOLBins are listed in Table 3-3.

Note 14

Microsoft Windows LOLBins are often categorized into binaries (programs that end in .EXE), libraries (.DLL), and scripts (.VBS). By some estimates, there are 115 different Windows LOLBins that can be exploited by a fileless virus while UNIX/ Linux systems have 185 LOLBins.

Table 3-3 Microsoft Windows common LOLBins

Name	Description
PowerShell	A cross-platform and open-source task automation and configuration management framework
Windows Management Instrumentation (WMI)	A Microsoft standard for accessing management information about devices
.NET Framework	A free, cross-platform, open-source developer platform for building different types of applications
Macro	A series of instructions that can be grouped together as a single command to automate a complex set of tasks or a repeated series of tasks and can be written by using a macro scripting language, such as Visual Basic for Applications (VBA), and is stored within the user document (such as in an Excel .xlsx worksheet or Word .docx file)

Unlike a file-based virus, a fileless virus does not infect a file and wait for that file to be launched. Instead, the malicious code of a fileless virus is loaded directly into the computer's random access memory (RAM) through the LOLBins and then executed.

Worm

A second type of malware that has as its primary purpose to spread is a worm. A **worm** is a malicious program that uses a computer network to replicate. (Worms are sometimes called *network viruses*.) A worm is designed to enter a computer through the network and then take advantage of a vulnerability in an application or an operating system on the computer. Once the worm has exploited the vulnerability of one system, it immediately searches for another computer on the network that has the same vulnerability.

Note 15

One of the first wide-scale worms occurred in 1988. This worm exploited a misconfiguration in a program that allowed commands emailed to a remote system to be executed on that system, and it also carried a payload that contained a program that attempted to determine user passwords. Almost 6,000 computers, or 10 percent of the devices connected to the Internet at that time, were affected. The threat actor who was responsible was later convicted of federal crimes in connection with this incident.

Early worms were relatively benign and designed simply to spread quickly but not corrupt the systems they infected. These worms slowed down the network through which they were transmitted by replicating so quickly that they consumed all network resources. Today's worms can leave behind a payload on the systems they infect and cause harm, much like a virus. Actions that worms have performed include deleting files on the computer or allowing the computer to be remotely controlled by an attacker.

Note 16

Although viruses and worms are said to be automatically self-replicating, *where* they replicate is different. A virus self-replicates *on* the host computer but does not spread to other computers by itself. A worm self-replicates *between* computers (from one computer to another).

Bot

Another popular payload of malware is software that allows the infected computer to be placed under the remote control of an attacker to launch attacks. This infected robot computer is known as a **bot** or *zombie*. When hundreds, thousands, or even millions of bot computers are gathered into a logical computer network, they create a *botnet* under the control of a *bot herder*.

Note 17

Due to the multitasking capabilities of modern computers, a computer can act as a bot while carrying out the tasks of its regular user. The user is completely unaware that his or her computer is being used for malicious activities.

Table 3-4 lists some of the attacks that can be generated through botnets.

Table 3-4 Uses of botnets

Type of attack	Description
Spamming	Botnets are widely recognized as the primary source of spam email. A botnet consisting of thousands of bots enables an attacker to send massive amounts of spam.
Spreading malware	Botnets can be used to spread malware and create new bots and botnets. Bots can download and execute a file sent by the attacker.
Ad fraud	Threat actors earn money by generating a high number of "clicks" on advertisements at targeted websites, using a bot to mimic the selections of a user.
Mining cryptocurrencies	Also called "cryptomining," this is a process in which transactions for various forms of cryptocurrency are verified, earning the "miner" a monetary reward. Botnets combine the resources of millions of bots for mining cryptocurrencies.

Infected bot computers receive instructions through a **command and control (C&C)** structure from the bot herders regarding which computers to attack and how. This communication occurs in a variety of ways, including the following:

- A bot can receive its instructions by automatically signing in to a website that the bot herder operates on which information has been placed that the bot knows how to interpret as commands.
- Bots can sign in to a third-party website; this has an advantage in that the bot herder does not need to have a direct affiliation with that website.
- Commands can be sent via blogs, specially coded attack commands through posts on Twitter, or notes posted on Facebook.
- Bot herders are increasingly using a "dead drop" C&C mechanism by creating a Google Gmail email account and then creating a draft email message that is never sent but contains commands that the bot receives when it logs in to Gmail and reads the draft. Because the email message is never sent, there is no record of it, and all Gmail transmissions are protected so that outsiders cannot view them.

Sidestep

The final category of malware is that which attempts to help malware "sidestep" or evade detection. This includes backdoor, logic bomb, and rootkit.

Backdoor

A **backdoor** gives access to a computer, program, or service that circumvents any normal security protections. Backdoors installed on a computer allow the attacker to return later and bypass security settings.

Creating a legitimate backdoor is a common practice by developers, who may need to access a program or device regularly, yet do not want to be hindered by continual requests for passwords or other security approvals. The intent is for the backdoor to be removed once the application is finalized. However, in some instances, backdoors have been left installed, and attackers have used them to bypass security.

Logic Bomb

A **logic bomb** is computer code that is typically added to a legitimate program but lies dormant and evades detection until a specific logical event triggers it. Once it is triggered, the program then deletes data or performs other malicious activities.

Logic bombs are difficult to detect before they are triggered. This is because logic bombs are often embedded in very large computer programs, some containing hundreds of thousands of lines of code. A trusted employee can easily insert a few lines of computer code into a long program without anyone detecting it. In addition, these programs are not routinely scanned for containing malicious actions.

Note 18

Many logic bombs have been planted by disgruntled employees. For example, a Maryland government employee tried to destroy the contents of more than 4,000 servers by planting a logic bomb script that was scheduled to activate 90 days after he was terminated.

Rootkits

A **rootkit** is malware that can hide its presence and the presence of other malware on the computer. It does this by accessing "lower layers" of the operating system or even using undocumented functions to make alterations. This enables the rootkit and any accompanying software to become undetectable by the operating system and common antimalware scanning software that is designed to seek and find malware.

Note 19

The risks of rootkits are significantly diminished today due to protections built into operating systems.

Two Rights & A Wrong

1. The two types of viruses are a file-based virus and a fileless virus.
2. A keylogger can be a software program or a small hardware device.
3. When hundreds, thousands, or even millions of bot computers are gathered into a logical computer network, they create a "swarm."

See Appendix A for the answer.

Computer Defenses

Given that there are many types of malware attacks against computers, and each of these types has many different instances, and new attacks are being continually introduced, how can a user today protect their computing device? The simple answer is no single defense can be implemented to protect a computer. Rather, it requires multiple types and layers of defenses.

A computer should have several security protections installed and configured to resist attacks. The defenses include managing patches, running antimalware software, examining firewalls, and stopping ransomware.

Managing Patches

Early versions of operating systems, such as Microsoft Windows, Apple macOS, and Linux, were simply "program loaders" whose job was to launch applications. As more features and graphical user interfaces (GUIs) were added, operating systems became more complex.

Note 20

Windows 11 is estimated to have over 50 million lines of code, Linux Debian has 68 million lines, and Apple macOS may have upward of 86 million lines of code. In contrast, Microsoft's first operating system, MS-DOS v1.0, had only 4,000 lines of code.

Due to the increased complexity of operating systems, unintentional vulnerabilities were introduced that could be exploited by attackers. In addition, new attack tools made vulnerable what were once considered secure functions and services on operating systems.

To address the vulnerabilities in operating systems that are uncovered after the software has been released to the general public as well as to provide ongoing additional features, operating system vendors deploy updates to users' computers through an automatic online update service. The user's computer interacts with the vendor's online update service to receive the latest updates.

The different names of the operating system updates that are provided by Microsoft for its Windows 11 product are listed in Table 3-5. However, more commonly, software updates to address a security issue are known as a **patch**.

Table 3-5 Microsoft operating system updates

Name	Release cycle	Description
Feature updates	Annually	These updates add new features to the operating system.
Quality updates	Second Tuesday of each month, although they can be released at any time	Quality updates deliver both security and non-security fixes and include security updates, critical updates, servicing stack updates, and driver updates.
Insider previews	Random	These "builds" are made available to interested users during the development process of new features that will be shipped in the next feature update.

Note 21

Microsoft quality updates are cumulative, so installing the latest quality update is sufficient to receive all the available patches for a specific Windows feature update.

The patch update options for Microsoft Windows 11 are shown in Figure 3-9. Clicking the **Check for updates** button will force the computer to connect to the Windows online update service to look for the latest updates. If updates are available, they can be installed by clicking the **Download & install** button. Under **More options**, additional update features are available. Updates can also be delayed from one to five weeks. After that time, the updates will resume. Other options include **Update history** that shows all of the updates installed and **Advanced options** that include setting a time range when the computer will restart after an update has been installed and optimizing delivery options that depend on the connection type.

Note 22

Before Windows 10, Microsoft users had several options regarding accepting or even rejecting patches. These options included *Install updates automatically, Download updates but let me choose whether to install them, Check for updates but let me choose whether to download and install them,* and *Never check for updates*. However, this approach frequently resulted in users ignoring important security patches and putting their computers at risk. More recent versions of Windows no longer have these options.

Figure 3-9 Windows 11 update options

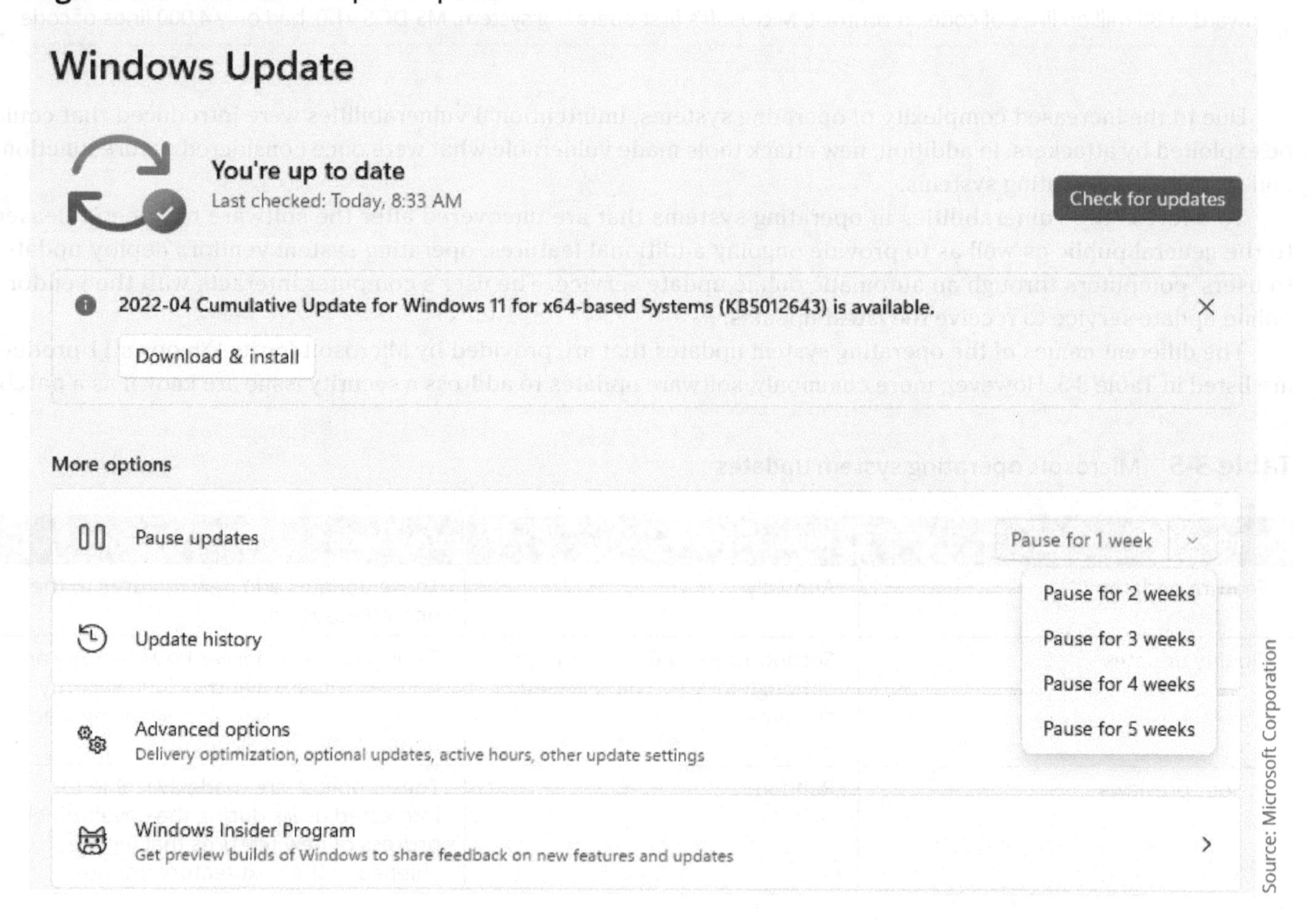

Source: Microsoft Corporation

Apple macOS has similar features. Users can receive a notification that software updates are available, can choose when to install the updates or choose to be reminded the next day. To check for macOS software updates, choose the **Apple** menu then **System Preferences** and click **Software Update**. To install updates, select **Install macOS updates**. To automatically download and install macOS updates, select **Automatically keep my Mac up to date**.

Promptly installing patches once they are available is *the most important step to protecting your computer.* It is advised that computers be configured so that patches are automatically downloaded and installed.

Note

A trend among software vendors is to automatically download and install patches without any user intervention or options. The Google Chrome web browser is automatically updated whenever necessary without even telling the user—and there are no user configuration settings to pause the updates.

Running Antimalware Software

Antimalware software is software that can combat various malware attacks. The most common antimalware software is antivirus (AV) software and comprehensive antimalware software.

Antivirus (AV) Software

One of the first antimalware software security applications was **antivirus (AV)** software. This software can examine a computer for any infections as well as monitor computer activity and scan new documents that might contain a virus. (This scanning is typically performed when files are opened, created, or closed.) If a virus is detected, options generally include cleaning the file of the virus, quarantining the infected file, or deleting the file.

AV software contains a virus-scanning engine and a database of known virus signatures, which are created by extracting a sequence of characters—a string—found in the virus that then serves as a virus' unique "signature." This database is called the *signature file*. By comparing the virus signatures against a potentially infected file, a match may indicate an infected file. The weakness of static analysis is that the AV vendor must constantly be searching for new viruses, extracting virus signatures, and distributing those updated databases to all users. Any out-of-date signature file could result in an infection.

At one time, running AV software was considered to be the primary—and often the only required—defense against attacks. However, due to the many types of malware, AV software, despite many users' perceptions, is no longer considered a "magic bullet" for providing complete ironclad protection on a computer.

Caution 

This false perception that AV software is the only solution needed to combat cybersecurity continues even today. Numerous studies have been conducted polling users regarding what cybersecurity defenses they use and recommend to others, and "Just install AV" is the most common solution that is given by users.

Comprehensive Antimalware Software

Modern antimalware software is more comprehensive than AV software which only looks for viruses. In addition to AV, comprehensive antimalware software often includes the following:

- **Intrusion prevention**. The software analyzes information arriving from a network and blocks potential threats before they enter a computer.
- **Reputation protection**. Using information gathered from a global network, it can classify software application files as "dangerous," "risky," or "safe" based on their attributes.
- **Behavioral protection**. This monitors applications for suspicious behavior and automatically blocks the software if necessary.

Both Microsoft Windows and Apple macOS have built-in comprehensive antimalware software, Microsoft Defender Antivirus and Apple XProtect. Figure 3-10 illustrates the Defender Antivirus options.

Figure 3-10 Microsoft Defender Antivirus options

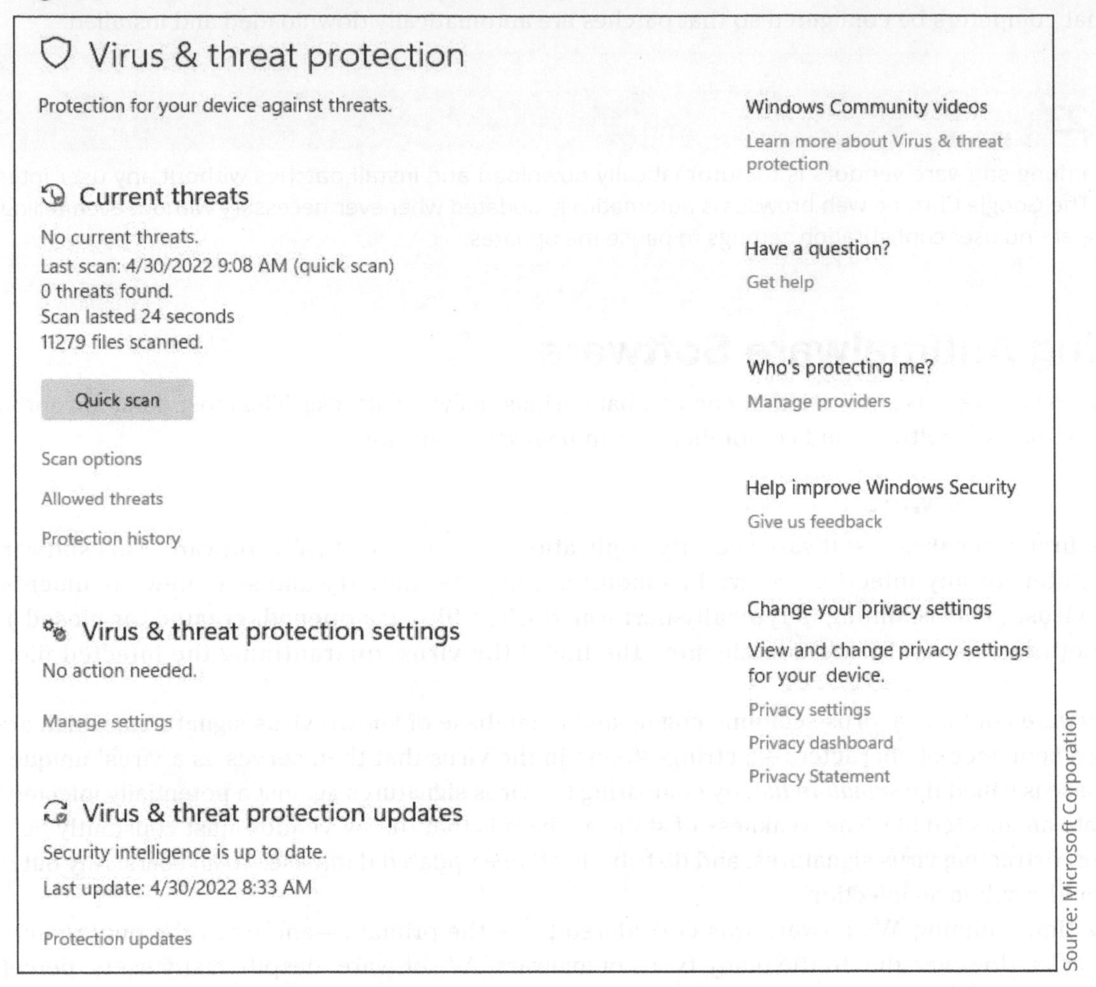

Source: Microsoft Corporation

Although antimalware software is not perfect and does not provide absolute protection, it does provide a degree of protection that would be missing if this software was not running. Without valid reasons, users should not turn off built-in comprehensive antimalware protection. However, it should be recognized as one tool in a large arsenal of weapons that must be deployed to defend against attackers.

Caution !

A wide variety of third-party comprehensive antimalware products can be purchased and installed on a computer. Experts debate whether these products actually do provide a significant degree of extra protection beyond the built-in products. Often these additional products may even conflict with the native products. Microsoft Defender does have an option to run in *Passive mode* instead of *Active mode*. Passive mode is used when it is not the primary antimalware software running on the computer: files are still scanned and detected threats are reported but they are not addressed by Defender.

Examining Firewalls

Both national and local building codes require commercial buildings, apartments, and other similar structures to have a *firewall*. In building construction, a firewall is usually a brick, concrete, or masonry unit positioned vertically through all stories of the building. Its purpose is to contain a fire and prevent it from spreading. A computer **firewall**, technically called a *packet filter*, serves a similar purpose: it is designed to limit the spread of malware.

Caution !

It is likely that no cybersecurity protection is more misunderstood than a firewall. Due to the nature of its name (*It's an impenetrable wall!*) and aided by inaccurate portrayals in movies and television, a firewall is often perceived by the general public as the ultimate security device that will block anything and everything malicious from entering a network. Unfortunately, this is a wildly inaccurate perception. Firewalls are an important element in cybersecurity, but they fall far short of being the ultimate defense.

There are two types of firewalls. A software-based **personal firewall** runs as a program on the local computer to block or filter traffic coming into and out of the computer. All modern operating systems include an *application firewall.* As the name suggests, these firewalls are application-oriented. An application or program running on a computer may need to communicate with another computer on the local network or an Internet server to send and receive information. The firewall would normally block these transmissions. With an application firewall, the user can create an "opening" in the firewall just for that program simply by approving the application to transmit (called *unblocking*). This is more secure than permanently opening an entry point on the firewall itself: when a permanent firewall opening is made, it always remains open and is then susceptible to attackers, but when an opening is unblocked by an application firewall, it is opened only when the application needs it. The settings for the Windows application firewall are shown in Figure 3-11.

Figure 3-11 Windows application firewall settings

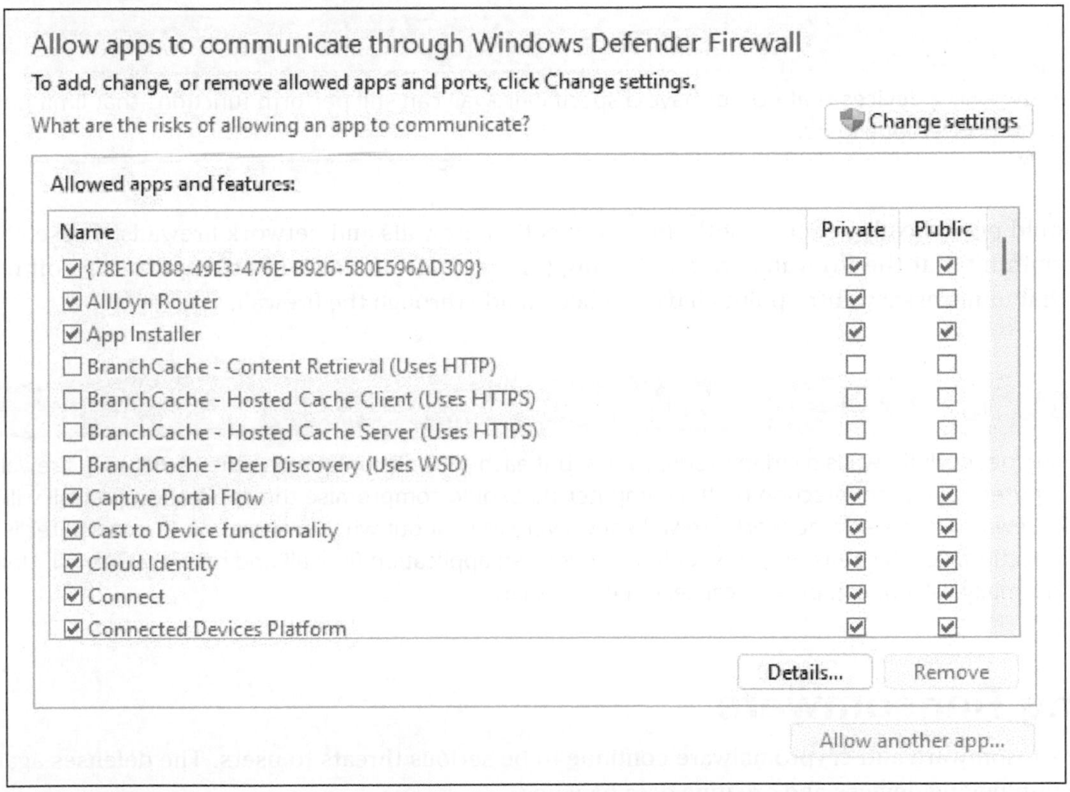

Allow apps to communicate through Windows Defender Firewall

To add, change, or remove allowed apps and ports, click Change settings.

What are the risks of allowing an app to communicate?

Change settings

Allowed apps and features:

Name	Private	Public
☑ {78E1CD88-49E3-476E-B926-580E596AD309}	☑	☑
☑ AllJoyn Router	☑	☐
☑ App Installer	☑	☑
☐ BranchCache - Content Retrieval (Uses HTTP)	☐	☐
☐ BranchCache - Hosted Cache Client (Uses HTTPS)	☐	☐
☐ BranchCache - Hosted Cache Server (Uses HTTPS)	☐	☐
☐ BranchCache - Peer Discovery (Uses WSD)	☐	☐
☑ Captive Portal Flow	☑	☑
☑ Cast to Device functionality	☑	☑
☑ Cloud Identity	☑	☑
☑ Connect	☑	☑
☑ Connected Devices Platform	☑	☑

Details... Remove

Allow another app...

Source: Microsoft Corporation

Note 24

Application firewalls can limit the spread of malware coming into the computer and prevent a user's infected computer from attacking other computers. For most application firewalls, inbound connections (data coming in from another source) are blocked unless a specific firewall rule setting allows them in. Outbound connections (data going out to another source) are allowed unless a specific rule blocks them and the outbound rules are turned on.

The second type of firewall is a hardware-based **network firewall**. Although an application software firewall that runs as a program on one computer is different in many respects from a network firewall designed to protect an entire network, their functions are essentially the same: to inspect network traffic and either accept or deny entry. Table 3-6 compares these two types of firewalls. Hardware firewalls are usually located at the "edge" of the network as the first line of defense defending the network and devices connected to it. Most home users have a network firewall as part of their networking equipment that provides Wi-Fi access or connects computers and devices such as printers together.

Table 3-6 Personal and network firewalls

Function	Personal firewall	Network firewall
Location	Runs on a single computer	Located on edge of the network
Scope of protection	Protects only the computer on which it is installed	Protects all devices connected to the network
Type	Software that runs on computer	Part of a Wi-Fi modem or separate hardware device
Filtering	Based on programs running on the computer	Provides a sophisticated range of filtering mechanisms

Note 25

Even home networking devices that do not have a specific firewall can still perform functions that limit entry into the network by unauthorized outsiders.

Users should periodically examine both their application firewalls and network firewalls. These checks should include determining that the firewalls are functioning (many types of malware attempt to turn off firewalls) and making sure that unnecessary entry points have not been made through the firewall.

Note 26

Application and network firewalls overlap in some ways, but each provides unique benefits. A network firewall is isolated from the computer so that an infection on the computer that could compromise the application firewall will not impact the network firewall. However, a network firewall knows very little about what program on the computer is making an outgoing connection. For this reason, you should have both an application firewall and network firewall. Unlike AV software, these normally will not compete or cancel out each other.

Stopping Ransomware

Both blocker ransomware and cryptomalware continue to be serious threats to users. The defenses against ransomware include unplugging devices and creating data backups.

Unplug Devices

Instead of ransomware only encrypting files on a local computer, today's ransomware will encrypt all files on *any* network or attached device connected to the endpoint computer. This includes secondary hard disk drives, USB hard drives, network-attached storage (NAS) devices, network servers, and even cloud data repositories.

To determine if a storage device could potentially be infected with ransomware, two tests can be applied. First, if a remote storage device is "mounted" on the local computer and can be freely accessed or displays a drive letter (like "D:"), then those files are at risk. Second, if a cloud storage repository is configured so that files automatically placed in a folder are then synced to the cloud storage, they too are at risk. This is because the ransomware can move encrypted files into the folder, and they will be replicated onto the cloud.

The solution is to "air gap" or physically isolate the computer from the storage device. Examples for different devices include the following:

- **External USB storage device**. Simply unplugging a USB storage device from the computer when it is not being used will protect it from ransomware.
- **Secondary hard disk drive**. A secondary hard drive can be "unmounted" and thus become invisible to the computer—and any ransomware—when the drive is not needed. It can then be mounted again when needed. (Unmounting a drive has no impact on the data stored in that drive.) For Microsoft Windows, enter the command **mountvol D: /p** at the command line or use the Windows Disk Management Utility to unmount a drive. For Apple macOS, in the Disk Utility app, select the disk and click the **Unmount** button.
- **Network-attached storage (NAS)**. For a network-attached storage (NAS) device, create a new share ("admin") and then create a new user account that is the only account that has access to it. Give the user account a strong username and password, and then log in (and out of) that share as needed.
- **Cloud storage**. It may be necessary to consider turning off automatic synchronization so that files placed in a folder are not immediately synced to your cloud storage. Instead, users should log into their cloud storage provider through a web browser that requires a username and password to sync the files. If this is not feasible, users can check with their cloud storage provider. Many provide some type of short-term "versioning," meaning that older versions are retained online for a limited period (perhaps seven days but sometimes up to a month). That means if the cloud storage files become encrypted, they can be rolled back to a previous version of unencrypted files.

Create Data Backups

One of the most important defenses against ransomware attacks as well as a wide range of malware is frequently overlooked: it is to create **data backups** regularly. Creating a data backup means copying files from a computer's hard drive onto other digital media that is stored in a secure location. Data backups protect against computer attacks because they can restore infected computers to their properly functioning state. Data backups can also protect against hardware malfunctions, user error, software corruption, and natural disasters.

There are several solutions to creating data backups. The two most common are continuous cloud backups and scheduled local backups.

Continuous Cloud Backups

The most comprehensive solution for most users is a *continuous cloud backup*. This backup occurs continually without any intervention by the user. Software monitors what files have changed and automatically updates the backed-up files with the most recent versions. These backups are stored online in the cloud. Several cloud-based services are available that provide features similar to the following:

- **Automatic continuous backup**. Once the initial backup is completed, any new or modified files are also backed up. Usually, the backup software will "sleep" while the computer is being used and perform backups only when there is no user activity. This helps to lessen any impact on the computer's performance or Internet speed.
- **Universal access**. Files backed up through online services can be made available to another computer.
- **Optional program file backup**. In addition to user data files, these services have an option to also back up all program and operating system files.
- **Delayed deletion**. Files that are copied to the online server will remain accessible for up to 30 days before they are deleted. This allows a user to have a longer window of opportunity to restore a deleted file.
- **Online or hardware-based restore**. If a file or the entire computer must be restored, this can be done online. Some services also provide the option of shipping to the user the backup files on a separate hardware device.

The advantages of online continuous backups are that they are performed automatically and stored at a remote location. These typically provide the highest degree of protection today for users.

Scheduled Local Backups A *scheduled local backup* is performed intentionally by the user and stored locally (not in the cloud). It could be performed every morning at 3:00 AM (automated) or whenever the user remembers that a backup is needed (on demand). When performing scheduled backups, several questions must be asked in advance to ensure the backup meets the users' needs.

The first question is *what* data should be backed up. All user-created files that cannot be easily or quickly recreated should be backed up. These include any personal files, such as word processing documents, digital photos, personal financial data, and other similar information. However, should programs installed on the computer, such as the operating system or a word processing program, also be backed up? Normally these programs are readily available elsewhere or can be retrieved easily, so there is little need to back them up along with the data files.

Note 27

The main reason to back up programs along with user data files is that it allows an infected computer to be completely restored more quickly from the backup instead of installing all of the programs individually before restoring the user data files.

The second question is *what media* should be used. A viable option is to use a portable USB hard drive. These devices connect to the USB port of a computer and provide backup capabilities; they are fast, portable, and can store large amounts of data.

The third question is *where to store* the backup. Consider a user who installs a second hard drive on his computer to back up the data from the primary hard drive each night. This would allow for the primary hard drive to be restored quickly in the event of a ransomware attack or even a hard drive failure. However, what about a fire, tornado, or lightning strike? These events could destroy both the primary hard drive and the backup hard drive. It is recommended that a copy of the data backup be stored offsite, such as at a work location or a friend's house.

Note 28

Home users should consider using the 3-2-1 backup plan. This plan says that you should always maintain *three* different copies of your backups (that does not count the original data itself) by using at least *two* different types of media on which to store these backups (a separate hard drive, an external hard drive, a USB device, online storage, etc.) and store *one* of the backups offsite.

The final question is *how frequently* the backup should be performed. It is recommended that backups be performed once per day on computers being used frequently. If that is not possible, then a regular schedule (such as every Tuesday and Friday) should be implemented and followed.

Note 29

Modern operating systems can perform automated backups, and third-party software is also available that provides additional functionality.

Two Rights & A Wrong

1. Installing AV is the most important step to protecting your computer.
2. It is recommended that users have both a personal firewall and some type of hardware firewall.
3. The most comprehensive backup solution for most users is a continuous cloud backup.

See Appendix A for the answer.

Module Summary

- Most successful attacks on computing devices are a result of malware. Malware (malicious software) is software that enters a computer system without the user's knowledge or consent and then performs an unwanted and harmful action. No standard has been established to classify the types of malware so that like malware can be grouped together for study. One attempt at classifying the diverse types of malware can be to examine the primary action that the malware performs and then group those together with similar actions. When classifying malware in this fashion, these malware actions used for groupings can be kidnap, eavesdrop, masquerade, serve as a launchpad, and sidestep.
- Some types of malware attacks perform a "kidnapping" on a user's computer and hold it "hostage" until a ransom is paid. Ransomware is the malicious software designed to extort money from victims in exchange for restoring their computer to its normal working state. There are two general types of ransomware. Blocker ransomware prevents the user from using their computer in a normal fashion. On occasion, blocker ransomware displays messages on the user's screen pretending to be from a reputable third party. Cryptomalware encrypts some or all the files on the device so that they cannot be opened until a ransom is paid. New variants of cryptomalware encrypt all files on the user's local hard drive along with any network or attached device connected to that computer.
- Another category of malware secretly spies on its victims. A keylogger silently captures and stores each keystroke that a user types on the computer's keyboard. An attacker can then search the captured text for useful information such as passwords, credit card numbers, or personal information. A keylogger can be a software program or a small hardware device. Spyware is tracking software that secretly monitors users by using the computer's resources, including programs already installed on the computer, to collect and distribute personal or sensitive information.
- Some malware attempts to deceive the user and hide its true intentions by pretending to be something else. A broad category of software that is often more annoying than malicious is called potentially unwanted programs (PUPs). A PUP is software that the user does not want on their computer. A computer Trojan is an executable program that masquerades as performing a benign activity but also does something malicious. A special type of Trojan is a remote access Trojan (RAT). A RAT has the basic functionality of a Trojan but also gives the threat agent unauthorized remote access to the victim's computer.
- Another category of malware infects a computer and then uses it to send attacks to other computers. A file-based virus is malicious computer code attached to a file. Each time the infected program is launched or the data file is opened, the virus first unloads a malicious payload. Then the virus reproduces itself by inserting its code into another file, but only on the same computer. A fileless virus does not attach itself to a file. Instead, fileless viruses take advantage of native services and processes that are part of the operating system to avoid detection and carry out their attacks. A worm uses a computer network to replicate. A worm is designed to enter a computer through the network and then take advantage of a vulnerability in an application or an operating system on the computer. Some malware allows the infected computer to be placed under the remote control of an attacker to launch attacks. This infected robot computer is known as a bot or zombie, and when multiple bot computers are gathered into a logical computer network, they create a botnet.
- Some malware attempts to evade detection. A backdoor gives access to a computer, program, or service that circumvents any normal security protections. A logic bomb is computer code that is typically added to a legitimate program but lies dormant and evades detection until a specific logical event triggers it. Once triggered, the program then deletes data or performs other malicious activities. A rootkit is malware that can hide its presence and the presence of other malware on the computer. It does this by accessing the lower layers of the operating system or undocumented functions to make alterations.
- No single defense can be implemented to protect a computer; it requires multiple types and layers of defenses. To address operating system vulnerabilities that are uncovered after the software has been released to the general public as well as to provide ongoing additional features, operating system vendors deploy updates to users' computers through an automatic online update service. Software updates to address a security issue are known as a patch. Promptly installing patches once they are available is the most important step in protecting a computer, so computers should be configured to automatically download and install patches.
- Antimalware software is software that can combat various malware attacks. One of the first antimalware software security applications was antivirus (AV) software that can examine a computer for any infections

as well as monitor computer activity and scan new documents that might contain a virus. If one is detected, options generally include cleaning the file of the virus, quarantining the infected file, or deleting the file. Modern antimalware software is more comprehensive than AV software, which only looks for viruses.

- A computer firewall is designed to limit the spread of malware. There are two types of firewalls. A software-based personal firewall runs as a program on the local computer to block or filter traffic coming into and out of the computer. All modern operating systems include an application firewall. An application firewall can provide an "opening" in the firewall just for that program. The second type of firewall is a hardware-based network firewall. These are usually located at the "edge" of the network as the first line of defense defending the network and devices connected to it. Most home users have a network firewall as part of their networking equipment that provides Wi-Fi access or connects computers and devices such as printers together.
- Instead of ransomware only encrypting files on a local computer, modern ransomware will encrypt all files on any network or attached device that is connected to the endpoint computer. The solution is to "air gap" or physically isolate the computer from the storage device. One of the most important defenses against ransomware and malware attacks is to create data backups regularly. Creating a data backup means copying files from a computer's hard drive onto other digital media that is stored in a secure location. The most comprehensive solution for most users is a continuous cloud backup that occurs continually without any intervention by the user. A scheduled local backup is performed intentionally by the user and stored locally.

Key Terms

antimalware software
antivirus (AV)
backdoor
blocker ransomware
bot
command and control (C&C)
cryptomalware
data backup
file-based virus
fileless virus
firewall
keylogger
logic bomb
malware
network firewall
patch
personal firewall
potentially unwanted program (PUP)
ransomware
remote access Trojan (RAT)
rootkit
spyware
Trojan
worm

Review Questions

1. Which of the following is NOT a reason why protecting your personal computer device is difficult?
 a. Many different types of attacks are launched against devices.
 b. Computing devices are too complex for average users to implement security.
 c. Attackers are constantly modifying their attacks.
 d. New attacks are created daily.

2. Which of the following requires a user to transport it from one computer to another?
 a. Adware
 b. Worm
 c. Virus
 d. Rootkit

3. Which of the following is NOT an action of a virus?
 a. Reformat the hard drive.
 b. Cause a computer to crash.
 c. Transport itself over the network.
 d. Erase files from a hard drive.

4. Which malware locks up a user's computer and then displays a message that purports to come from a law enforcement agency?
 a. Trojan
 b. Cryptomalware
 c. Ransomware
 d. Worm

5. Which of the following is NOT a type of malware that has as its primary trait to launch attacks on other computers?
 a. Bot
 b. Virus
 c. Worm
 d. Trojan

6. Which of the following is NOT a type of malware?
 a. Kidnap
 b. Sidestep
 c. Diffuse
 d. Eavesdrop

7. **Rowan's sister called him about a message that suddenly appeared on her screen that says her software license has expired and she must immediately pay $500 to have it renewed before control of the computer will be returned to her. What type of malware has infected her computer?**

- **a.** Persistent lockware
- **b.** Blocking ransomware
- **c.** Cryptomalware
- **d.** Impede-ware

8. **Marius' team leader has just texted him that a supervisor has an employee who violated company policy by bringing in a file on her USB flash drive and her computer is suddenly locked up with cryptomalware. Why would Marius consider this a dangerous situation?**

- **a.** It sets a precedent by encouraging other employees to violate company policy.
- **b.** Cryptomalware can encrypt all files on any network that is connected to the employee's computer.
- **c.** The organization may be forced to pay up to $500 for the ransom.
- **d.** The employee would have to wait at least an hour before her computer could be restored.

9. **Which malware takes advantage of native services and processes that are part of the operating system to avoid detection and carry out attacks?**

- **a.** PUP
- **b.** File-based virus
- **c.** Fileless virus
- **d.** NET-Bot

10. **What type of program has Fleur installed that prints coupons but in the background silently collects her passwords?**

- **a.** Backdoor
- **b.** Rootkit
- **c.** Remote Access Trojan (RAT)
- **d.** Trojan

11. **Which of the following is known as a network virus?**

- **a.** TAR
- **b.** Worm
- **c.** Remote exploitation virus (REV)
- **d.** C&C

12. **Uwe is researching the different types of attacks that can be generated through a botnet. Which of the following would NOT be something distributed by a botnet?**

- **a.** Cracker
- **b.** Spam
- **c.** Malware
- **d.** Ad fraud

13. **Which of the following is NOT a means by which a bot communicates with a C&C device?**

- **a.** Signing into a website the bot herder operates
- **b.** Signing into a third-party website
- **c.** Email
- **d.** Command sent through Twitter posts

14. **Jurgen's brother is complaining to him about all of the software on his new computer that came preinstalled but that he does not want. What type of software is this?**

- **a.** Spyware
- **b.** Bot
- **c.** PUP
- **d.** Keylogger

15. **What is the difference between a Trojan and a RAT?**

- **a.** There is no difference.
- **b.** A RAT gives the attacker unauthorized remote access to the victim's computer.
- **c.** A RAT is primarily used to carry malware while a Trojan does not.
- **d.** A RAT can only infect a smartphone and not a computer.

16. **Which of these would NOT be considered the result of a logic bomb?**

- **a.** Send an email to Han's inbox each Monday morning with the agenda of that week's department meeting.
- **b.** If the company's stock price drops below $50, then credit Oscar's retirement account with 1 additional year of retirement credit.
- **c.** Erase the hard drives of all the servers 90 days after Alfredo's name is removed from the list of current employees.
- **d.** Delete all human resource records regarding Augustine one month after he leaves the company.

17. **Which of these is a general term used for describing software that gathers information without the user's consent?**

- **a.** Pullware
- **b.** Spyware
- **c.** Scrapeware
- **d.** Adware

18. **What is the general name for a software fix that addresses a security vulnerability?**

- **a.** Security update
- **b.** Feature pack
- **c.** Security plug
- **d.** Patch

19. Which of the following is a defense against ransomware?

- **a.** Create data backups.
- **b.** Check for a rootkit each week.
- **c.** Look for a backdoor in a computer.
- **d.** Install fileless virus protection.

20. Which statement regarding a keylogger is NOT true?

- **a.** Software keyloggers can be designed to send captured information automatically back to the attacker through the Internet.
- **b.** Hardware keyloggers are installed between the keyboard connector and computer keyboard USB port.
- **c.** Software keyloggers are generally easy to detect.
- **d.** Keyloggers can be used to capture passwords, credit card numbers, or personal information.

Hands-On Projects

Project 3-1: Downloading and Running the Microsoft Safety Scanner

Could your computer already have malware on it and you don't know it? Microsoft offers a free tool that can scan a computer for malware and then remove it. In this project, you will download and run the Microsoft Safety Scanner.

1. Open your web browser and enter the URL **docs.microsoft.com/en-us/microsoft-365/security/intelligence/safety-scanner-download**. (If you are not able to access the site through the web address, use a search engine to search for "Microsoft Safety Scanner Download.")
2. Read the information on this webpage about the scanner.
3. Click the **Download Microsoft Safety Scanner (64-bit)** link.
4. After the file downloads, open the file by double-clicking it.

Caution !

Note where you saved this downloaded file. The Microsoft Safety Scanner is a portable executable and will not appear in the Windows Start menu or as an icon on the desktop.

5. Answer **Yes** if asked **do you want to allow this app to make changes to your device?**
6. When the End user license agreement appears, click the box **Accept all terms of the preceding license agreement** and then click **Next**.
7. When the **Welcome to the Microsoft Safety Scanner** window appears, click **Next**.
8. When the **Scan type** window appears, if necessary, click **Quick scan**. Click **Next**.
   The Microsoft Safety Scanner will perform a limited scan of the computer. This scan may take up to 5 minutes to complete depending on the computer's configuration.
9. A **Scan results** window appears when the scan is completed. If it has detected any suspicious files or malware, you may be asked to conduct a full scan.
10. How easy was this scanner to use? Is this something that you would recommend to a friend? Why or why not?

Note 30

The Microsoft Safety Scanner can be downloaded and executed at any time. It only scans when manually launched by a user. A downloaded version is available for use up to 10 days after being downloaded; after this, it expires. It is recommended that users always download the latest version of this tool before each scan.

11. Click **Finish**.
12. Close all windows.

Project 3-2: Configuring Microsoft Windows Security – Part 1

Security settings must be properly configured on a computer to protect it. In this project, you will examine several security settings on a Microsoft Windows 11 computer using the Windows interface.

Caution !

This project shows how to configure Windows security for a personal computer. If this computer is part of a computer lab or office, these settings should not be changed without the proper permissions.

1. Click **Start** and then click **Settings**.
2. Click **Privacy & security**.
3. Click **Windows Security**.
4. Note the Protection areas and if any actions are necessary.
5. Click **Open Windows Security**.
6. Click **Virus & threat protection**. Note that a Quick scan of the computer can also be performed from this window.
7. Under **Virus & threat protection settings**, click **Manage settings**.
8. Be sure that all of the options are set to On as shown in Figure 3-12.

Figure 3-12 Virus & threat protection settings

Virus & threat protection settings

View and update Virus & threat protection settings for Microsoft Defender Antivirus.

Real-time protection

Locates and stops malware from installing or running on your device. You can turn off this setting for a short time before it turns back on automatically.

On

Cloud-delivered protection

Provides increased and faster protection with access to the latest protection data in the cloud. Works best with Automatic sample submission turned on.

On

Automatic sample submission

Send sample files to Microsoft to help protect you and others from potential threats. We'll prompt you if the file we need is likely to contain personal information.

On

Submit a sample manually

Tamper Protection

Prevents others from tampering with important security features.

On

Source: Microsoft Corporation

Continued

9. Click the **Back** button to return to the **Virus & threat protection** screen.
10. Under **Virus & threat protection updates**, click **Protection updates**.
11. Click **Check for updates** to be sure that the latest updates have been downloaded and applied.
12. Click the **Back** button to return to the **Virus & threat protection** screen.
13. Click the **Back** button to return to the **Security at a glance** screen.
14. Click **Firewall & network protection** in the left pane.
15. In the **Firewall & network protection** screen, be sure that all of the networks listed have **Firewall is on** below the network.
16. Click **Allow an app through firewall**.
17. Scroll down through the **Allowed apps and features** list. It is here that an app would be turned on or off to allow it to communicate through the firewall. To add an app not listed, you would click **Change settings** and then click **Allow another app**.
18. Close the **Allowed apps** window.
19. Click the **Back** button on the **Firewall & network protection** screen to return to the **Security at a glance** screen.
20. Close the **Security at a glance** screen to return to the **Privacy & security > Windows Security** screen.
21. Remain here for the next project.

Project 3-3: Configuring Microsoft Windows Security – Part 2

In this project, you will continue to examine several security settings on a Microsoft Windows 11 computer using the Windows interface.

Caution !

This project shows how to configure Windows security for a personal computer. If this computer is part of a computer lab or office, these settings should not be changed without the proper permissions.

1. Click **Windows Update** to display the screen for managing patches.
2. If any updates need to be installed, click **Install now**.
3. Under **More options**, display the **Pause updates** menu to pause downloading updates. How many options are available to pause? What would be a reason why someone would pause these updates? Set your time to pause to the minimum setting.
4. Click the right arrow next to **Update history**.
5. View the updates that have been already installed. How many are there? What are the dates for these updates? How frequently have they occurred?
6. Scroll down to **Definition Updates**. These are the virus signature updates for the native Microsoft Defender Antivirus software.
7. Click the **Back** arrow to return to the **Windows Update** screen.
8. Click the **Back** arrow to return to the **Privacy & security > Windows Security** screen.
9. Click **Open Windows Security**.
10. Click **Virus & threat protection**.
11. Under **Ransomware protection**, click **Manage ransomware protection.**
    Microsoft Windows Defender offers built-in protection against ransomware. This is called *Controlled folder access*. This means that only trusted apps can access the content in protected folders, thus denying an attempted cryptomalware attack from encrypting documents in these folders. These protected folders are set by default and include those folders that typically contain user data, such as Documents, Pictures, Videos, Music, and Favorites. You can also add other folders to protect and other approved apps to access protected folders. In the event of a ransomware attack, a notification appears with a warning that an app attempted to make changes to a file in a protected folder.
12. Turn on **Controlled folder access** and, if necessary, click **Yes**.

13. Click **Protected folders** and, if necessary, click **Yes** to see the folders that are now protected.
14. How valuable is Controlled folder access? How easy is it to use for protection against ransomware?
15. Close all windows.

Project 3-4: Configuring Microsoft Windows Security – Part 3

As seen from the prior projects, Windows security settings are found across several screens. This can make it easy to overlook important settings and time-consuming to fine-tune the settings, especially when configuring the Microsoft Defender virus and threat protection product. A third-party tool called ConfigureDefender provides an easier interface. In this project, you will download and use the ConfigureDefender product.

Note 31

ConfigureDefender is not installed on the computer but runs as a stand-along application.

1. Open your web browser and enter the URL **https://github.com/AndyFul/ConfigureDefender**. (If you are no longer able to access the program through the URL, use a search engine to search for "ConfigureDefender.")
2. Click **ConfigureDefender.exe**.
3. Click **Download**.
4. Open the file after the download has been completed.
5. Answer **Yes** if asked **do you want to allow this app to make changes to your device?**
6. The ConfigureDefender app now launches, as shown in Figure 3-13.

Figure 3-13 ConfigureDefender

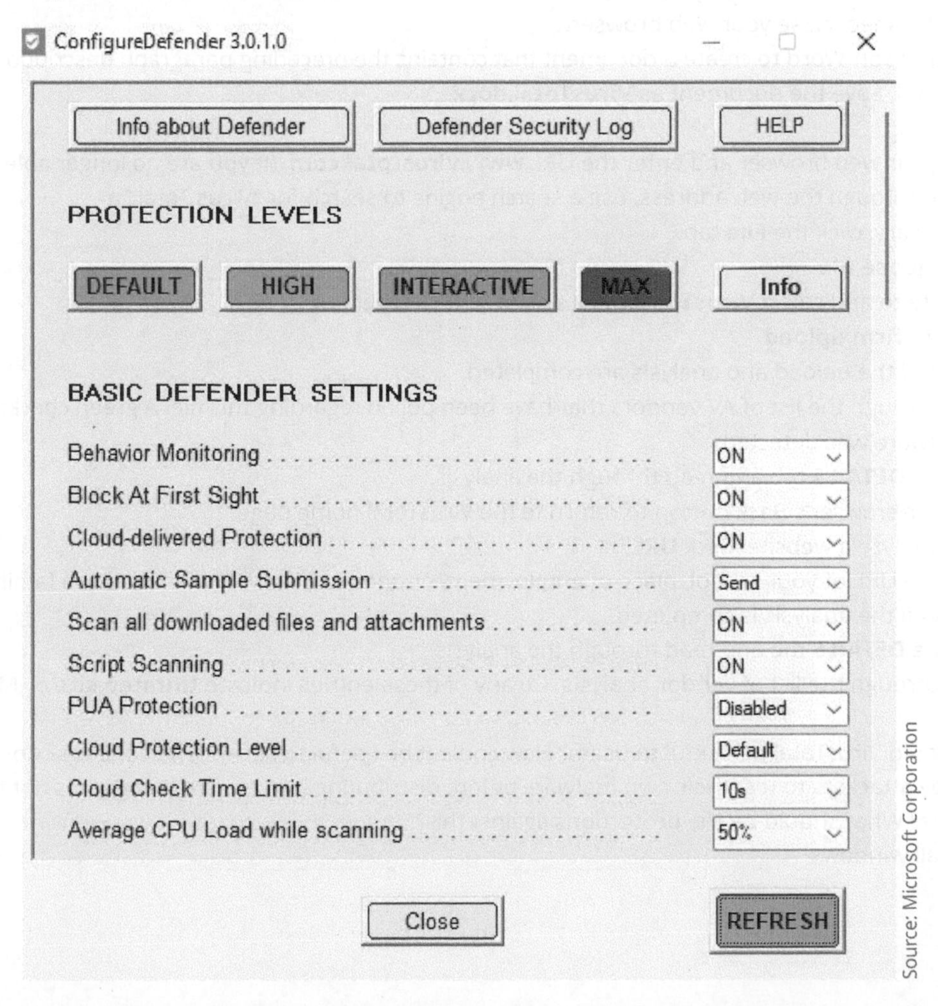

Source: Microsoft Corporation

Continued

7. Scroll down through the settings. Were you aware that Windows Defender has so many options?
8. Click the **Info** button to see the different protection levels. Close this window when finished reading.
9. Click the **HIGH** button and then close the pop-up box.
10. Scroll down through the settings. How much stronger are they than the Default settings?
11. Click the **MAX** button and then close the pop-up box.
12. Scroll down through the settings. How much stronger are they?
13. Finally, click either the **DEFAULT** or **HIGH** button to set your computer to the security level that you choose, or click **DEFAULT** to return to the basic security level.
14. How easy is ConfigureDefender to use? Would you recommend it to others?
15. Close all windows.

Project 3-5: Analyze Files and URLs for File-Based Viruses Using VirusTotal

VirusTotal is a free online service that analyzes files and URLs to identify potential malware. VirusTotal combines 70 antivirus scanners and URL/domain blacklisting services along with other tools to identify malware. A wide range of files can be submitted to VirusTotal for examination, such as user data files and documents, executable programs, PDFs, and images. One use of VirusTotal is to provide a "second opinion" on a file or URL that may have been flagged as suspicious by other scanning software. In this project, you use VirusTotal to scan a file and a URL.

1. First view several viruses from 20 years ago and observe their benign but annoying impact. Open your web browser and enter the URL **https://archive.org/details/malwaremuseum&tab=collection**. (If you are no longer able to access the site through the web address, use a search engine to search for "Malware Museum.")
2. All of the viruses have been rendered ineffective and will not harm a computer. Click several of the viruses and notice what they do.
3. When finished, close your web browser.
4. Use Microsoft Word to create a document that contains the preceding paragraph description about VirusTotal. Save the document as **VirusTotal.docx**.
5. Exit Word.
6. Open your web browser and enter the URL **www.virustotal.com**. (If you are no longer able to access the site through the web address, use a search engine to search for "Virus Total.")
7. If necessary, click the **File** tab.
8. Click **Choose file**.
9. Navigate to and select **VirusTotal.docx** and then click **Open**.
10. Click **Confirm upload**.
11. Wait until the upload and analysis are completed.
12. Scroll through the list of AV vendors that have been polled regarding this file. A green checkmark means no malware was detected.
13. Click the **DETAILS** tab and read through the analysis.
14. Use your browser's Back button to return to the VirusTotal home page.
15. Next, analyze a website. Click **URL**.
16. Enter the URL of your school, place of employment, or another site with which you are familiar.
17. Wait until the analysis is completed.
18. Click the **DETAILS** tab and read through the analysis.
19. Scroll through the list of vendor analysis. Do any of these entries indicate **Unrated site** or **Malware site**?
20. How could VirusTotal be useful to users? How could it be useful to security researchers? Could it also be used by attackers to test their own malware before distributing it to ensure that it does not trigger an AV alert? What should be the protections against this?
21. Close all windows.

Project 3-6: Explore Ransomware Information and Tool Sites

A variety of sites provide information about ransomware along with tools for counteracting some types of infection. In this project, you explore different ransomware sites.

1. Open your web browser and enter the URL **www.nomoreransom.org**. (If you are not able to access this site, open a search engine and search for "Nomoreransom.org.")
2. Click the **No** button.
3. Read through the Prevention Advice. Do you think it is helpful?
4. Click **Crypto Sheriff**. How could this be useful to a user who has suffered a ransomware infection?
5. Click **Ransomware: Q&A**. Read through the information. Which statements would you agree with? Which statements would you disagree with?
6. Click **Decryption Tools**. These tools may help restore a computer that has been infected by a specific type of ransomware.
7. Click one of the tools and then click **Download** to download it. Note that these tools change frequently based on the latest types of ransomware that are circulating.
8. Run the program to understand how these decryption tools work. Note that you will not be able to complete the process because there are no encrypted files on the computer. Close the program.
9. Now visit another site that provides ransomware information and tools. Open your web browser and enter the URL **https://id-ransomware.malwarehunterteam.com/**.
10. What features does this site provide?
11. How could these sites be useful?
12. Close all windows.

Case Projects

Case Project 3-1: Biological and File-Based Viruses

The word virus comes from Latin, meaning a slimy liquid, poison, or poisonous secretion. In late Middle English, it was used for the venom of a snake. The word later evolved from the discharge to the substances within the body that caused the infectious diseases that produced the discharge. In 1799, Edward Jenner published his discovery that the "cow-pox virus" could actually be used as a vaccine against smallpox. As biological science continued to advance, the word "virus" became even more specific when referring to tiny infectious agents—even smaller than bacteria—that replicate in living cells. This new field of "virology" exploded in the 1930s when electron microscopes allowed scientists to see viruses for the very first time. Since then, scientists have continued to identify and name new biological viruses. Combating viruses by developing vaccines has many parallels to how malicious file-based viruses are identified and removed from a computer. Using the Internet, research these two types of viruses and find the similarities between combating biological and computer viruses. Write a one- to two-paragraph summary of your research.

Case Project 3-2: Reasons for Zero-Day Attacks

Vulnerabilities occur in software that result in zero-day attacks for many reasons. One reason is improper testing being conducted before the software is released that could identify the vulnerability and correct it. Another reason is most programmers who write code simply do not think like an attacker, so even when they try to be secure, they only address what is obvious to them. What are some other reasons? Use the Internet to research reasons why security vulnerabilities in software occur. Identify at least four reasons. Then, give a short explanation about how these could be addressed to reduce the number of vulnerabilities that create zero-day attacks. Write a one-page paper on your research.

Continued

Case Project 3-3: Infamous Logic Bombs

Search the Internet for examples of logic bombs. Select three logic bombs and write a report about them. Who was responsible? When did the bombs go off? What was the damage? What was the penalty for the person responsible? Did the organization make any changes after the attack? Is there any way to prevent logic bombs? Write a one-page paper about what you have learned.

Case Project 3-4: Banning Ransomware Payments

Different state legislatures are now considering legislation that would ban any state or local government agencies from paying a ransom. The reasoning is that if the attackers know that a state or local government is prohibited from paying a ransom, then these entities will no longer be victims of attacks since the attackers know they will never be paid. It is also said that the laws would also decrease the total level of ransomware, since paying ransoms only encourages more of this malicious activity. Some have suggested that any ransomware payments by any business or organization should be deemed illegal. But will this actually stop ransomware? What would happen to those entities that are hit with ransomware and are unable to restore their systems from backups? Debate the pros and cons of banning ransomware payments and write a one-page paper about these two sides. What is your recommendation? Why?

Case Project 3-5: Lines of Computer Code

How do different software applications compare in terms of the number of lines of code that they have? Visit the "Codebases: Millions of lines of code" on the Information is Beautiful website (www.informationisbeautiful.net/visualizations/million-lines-of-code/). Compare the length of different applications. Notice that the gray semicircles connect a previous version of an application with a subsequent version to illustrate how modern programs have evolved. Search for applications that you currently use or have used in the past. Does the length of these programs surprise you? Does this make it easier to understand how vulnerabilities can creep into computer code? Can you make any suggestions regarding how to make computer code more secure? Write a one-page paper on your analysis.

References

1. "McAfee Labs Threats Report," Dec. 2018. Accessed Apr. 21, 2019. www.mcafee.com/enterprise/en-us/assets/reports/rp-quarterly-threats-dec-2018.pdf.
2. Kerner, Sean Michael, "Ransomware trends, statistics and facts in 2022," *TechTarget*, Feb. 2022. Accessed Apr. 29, 2022. https://www.techtarget.com/searchsecurity/feature/Ransomware-trends-statistics-and-facts#:~:text=This%20represents%20a%2062%25%20year,16%20U.S.%20critical%20infrastructure%20sectors.
3. "Ransomware attacks map," *Statescoop*. Accessed Apr. 29, 2022. https://statescoop.com/ransomware-map/.
4. "Ransomware payments up 33% as Maze and Sodinokibi proliferate in Q1 2020," *Coveware*. Accessed May 5, 2020. https://www.coveware.com/blog/q1-2020-ransomware-marketplace-report.
5. "High-impact ransomware attacks threaten U.S. businesses and organizations," *Public Service Announcement Federal Bureau of Investigation*, Oct. 2, 2019. Accessed May 9, 2020. https://www.ic3.gov/media/2019/191002.aspx.

Module 4

Internet Security

After completing this module, you should be able to do the following:

1. Explain how the World Wide Web and email work.
2. Identify the risks associated with using a browser and email.
3. Explain the threats from web servers and transmissions.
4. Describe the steps in securing a web browser.
5. List email defenses.

Cybersecurity Headlines

Suppose after attending a class you notice a wallet on the floor that was evidently dropped by another student. You open the wallet and see the driver's license with the student's name and recognize that the student was indeed in the class. As you start to take the wallet to the lost and found, you see the panicked student in the hallway. You go over to him and hand to him his wallet. But instead of thanking you, the student becomes livid and claims that you have violated his privacy by looking inside the wallet at his driver's license. He then goes on to say that he will sue you for a violation of privacy. As the student snatches the wallet from your hand and storms off, what are you thinking? Many of us would be tempted to think, "Well, no good deed goes unpunished!"

That may have been exactly what went through the mind of a journalist recently. After discovering a massive data leak just by using a web browser, the journalist responsibly reported it so that the leak could be fixed. However, the governor of that state then threatened to prosecute the journalist and bring criminal charges. The governor went on to say the state would also seek civil damages, claiming that the journalist is a "hacker," and that the newspaper's reporting was just a "political vendetta" and "an attempt to embarrass the state and sell headlines for their news outlet."

Josh Renaud, a *St. Louis Post-Dispatch* web developer and journalist, was looking at a website maintained by the Missouri Department of Elementary and Secondary Education (DESE), which listed teacher credentials and certifications. When he displayed the source code of the webpage (as easy as pressing *Ctrl+u* in a Google Chrome web browser) he was shocked to find that the Social Security numbers (SSNs) of over 100,000 teachers, counselors, and even school administrators could be seen. Renaud did not have permission from the state of Missouri to probe the depth of the website's security. Rather, he had just stumbled upon the exposed data through his web browser.

Renaud took the responsible step of privately alerting the DESE so they could fix the problem with this website—and allow the state to examine similar web applications and also fix them—before he published information about it. The news report was published one day after the department removed the affected pages from its website.

That's when Mike Parson, the governor of Missouri, jumped in. Parson said, *"Nothing on DESE's website gave permission or authorization for this individual to access teacher data. This individual is not a victim. They were acting against a state agency to compromise teachers' personal information in an attempt to embarrass the state and sell headlines for*

Continued

their news outlet. We will not let this crime against Missouri teachers go unpunished, and we refuse to let them be a pawn in the news outlet's political vendetta. Not only are we going to hold this individual accountable, but we will also be holding accountable all those who aided this individual and the media corporation that employs them." The governor went on to claim that this incident may cost Missouri taxpayers up to $50 million while diverting workers and resources from other state agencies.

Is what Renaud did wrong?

The Computer Fraud and Abuse Act (CFAA) was enacted in 1986 as an amendment to the first federal computer fraud law. Since then, it has been amended multiple times (most recently in 2008) to now cover a broad range of conduct. The CFAA prohibits intentionally accessing a computer without authorization or in excess of authorization. But it fails to define what "without authorization" means.

Is looking at the source code of a webpage, as Renaud did, intentionally accessing a computer "without authorization"? It can be argued that the website source code is freely provided to all web browsers (otherwise the browser could not display the webpage), so the argument that Renaud was "intentionally accessing a computer without authorization" seems to be without merit.

The governor of Missouri also said that Renaud can be prosecuted under Missouri Code 569.095, which defines the offense as "tampering with computer data if he or she knowingly and without authorization or reasonable grounds to believe that he has such authorization." Again, could Renaud argue that he had the authorization to look at the source code of the website since it was provided to him from the web server? This Missouri code further defines the offense as an act by someone who "Accesses a computer, a computer system, or a computer network, and intentionally examines information about another person." Did Renaud access another computer, or was the webpage given to his computer from the web server?

After two months of investigation by the Missouri State Highway Patrol, a police report was issued. The report confirmed that Renaud did exactly what he had said from the beginning: he identified a security flaw by viewing publicly available information on a misconfigured state website and delayed publishing an article on his findings until after the state closed the security vulnerability. The report also revealed that the security flaw was more serious than first thought. The vulnerability had existed since 2011, up to 576,000 teachers' SSNs may have been exposed, and the exposed data goes as far back as 2005.

After an additional two months of investigation after the police report was first issued, the county prosecutor finally weighed in. He declined to bring any charges against Renaud. The investigation was then closed.

The impact of the Internet on our world has been nothing short of astonishing. Although today's Internet has its roots back in the late 1960s, for almost a quarter of a century it was used only by researchers and the military. With the introduction of web browser software in the early 1990s and the spread of telecommunication connections at work and home, the Internet became useable and accessible to virtually everyone. This created a seismic shift across all societies around the world. First, a virtually limitless amount of information became freely available at users' fingertips. Second, the Internet created a collective force of tremendous proportions. For the first time in human history, mass participation and cooperation across space and time became possible, empowering individuals and groups all over the world. The Internet is truly having a revolutionary impact on how we live.

But for all of the benefits that the Internet has provided, it also has become the primary pathway for threat actors to unleash their attacks against us. When we connect a device to the Internet, we also are exposing that device to malicious attacks. This can be illustrated through a recent event conducted by security researchers. These researchers installed 10 *honeypots* on the Internet. A honeypot is a computer located in an area with limited security to serve as "bait" to threat actors so that any new attacks against the honeypot could be analyzed. In this event, one of the honeypots started receiving login attempts from attackers just *52 seconds* after it went online. After all 10 of the honeypots were discovered by attackers, a login attempt was made about every *15 seconds* on the honeypots. At the end of one month, over *5 million attacks* had been attempted on these honeypots.[1]

In this module, you will learn about how the Internet works and then identify the types of risks with using it. You will then examine the defenses that can be set up to make it a safer environment when using it.

How the Internet Works

The **Internet** is a global computer network that allows connected devices to exchange information. Yet there are two common misconceptions regarding the Internet. First, the Internet is not made up of individual devices (desktop computers, tablets, laptops, smartphones, etc.) but instead is composed of networks around the world to which devices are attached. In fact, the Internet is often defined as *an international network of computer networks*. Second, it is not owned or controlled by any single organization or government entity. Instead, these networks are operated by industry, governments, schools, and even individuals, who all loosely cooperate to make the Internet a global power.

Understanding how some of the basic Internet tools work helps to provide the foundation for establishing Internet security. Two of the primary Internet tools that are used today are the World Wide Web and email. It is no coincidence that through these tools the overwhelming majority of Internet attacks occur.

The World Wide Web

The **World Wide Web (WWW)**, better known as just the *web*, is composed of Internet server computers on networks that provide online information in a specific format. The format of the information is based on the **Hypertext Markup Language (HTML)**. HTML allows web authors to combine text, graphic images, audio, video, and **hyperlinks** (so users may jump from one area on the web to another with just a click of the mouse button). Instructions written in HTML code specify how the words, pictures, and other elements should be displayed on a user's screen. HTML code is combined into a series of *webpages* that make up a *website*, much like how one book is composed of multiple yet individual pages. A sample of a webpage and the HTML code behind it is shown in Figure 4-1.

Figure 4-1 HTML code

```
<!doctype html>
<html class="mod_no-js" lang="en" ng-app="app">
<head>
  <meta charset="utf-8">
  <meta name="viewport" content="width=device-width, initial-scale=1">
<title>Cengage: Digital Course Solutions &amp; Online Textbooks - Cengage</title>
<script>
//<![CDATA[

    window.components = window.components || {};
    window.components.seo = {
      cafeCategory: 'Digital Learning & Online Textbooks'
    }

//]]>
</script>
<meta name="description" content="Higher education course materials–like college textbooks, eTextbooks and online learning platforms–to help make the things you want to do possible."/>
<meta name="keywords" content="Cengage, digital learning, online textbooks, learning materials, online learning"/>
<meta property="og:type" content="article"/>
<meta property="og:title" content="Cengage: Digital Course Solutions &amp; Online Textbooks"/>
<meta property="og:image" content="https://embed.widencdn.net/img/cengage/ukotlxykzh/400px@1x/weblogo-cengage.ai.png?q=100"/>
<meta property="og:description" content="Higher education course materials–like college textbooks, eTextbooks and online learning platforms–to help make the things you want to do possible."/>
<link rel="canonical" href="https://www.cengage.com/"/>
```

The HTML code on a webpage is interpreted and the resulting content is displayed through software running on a user's device. This software is called a **web browser**. A web browser displays the words, pictures, and other elements following the instructions given through the HTML code for that webpage.

Note 1

Before web browsers, users were forced to type lengthy and complicated commands to navigate the Internet, which discouraged its widespread use. The first web browser was introduced in 1990 by Tim Berners-Lee, who was the inspiration behind the web itself. He called his browser the *WorldWideWeb* but later renamed it *Nexus*. At that time, it was the only browser available and thus the only way to access the web. In 1993, Mosaic became the first browser to allow images embedded in text, followed one year later by Netscape Navigator, which quickly became the dominant web browser for the remainder of the decade. Today, Google Chrome, Apple Safari, Mozilla Firefox, and Microsoft Edge are the most popular browsers.

Web servers distribute HTML documents based on a set of standards, or *protocols*, known as the **Hypertext Transfer Protocol (HTTP)**. HTTP is a subset of a larger set of standards for Internet transmission known as the **Transmission Control Protocol/Internet Protocol (TCP/IP)**.

Note 2

The word *protocol* comes from two Greek words for *first* and *glue*, and originally referred to the first sheet glued onto a manuscript on which the table of contents was written. The term later evolved to mean an "official account of a diplomatic document" and was used in France to refer to a formula of diplomatic etiquette.

Figure 4-2 illustrates how a webpage is displayed based on a user's request. The user opens the web browser on her smartphone and enters a **uniform resource locator (URL)**, which is a web address such as www.cengage.com. The web browser sends a request to the remote web server using HTTP. The web server then sends the requested HTML document (again using HTTP), which is stored on the user's smartphone. Finally, the user's web browser, using the code contained in the HTML document, displays the results on the web browser.

Figure 4-2 Webpage display

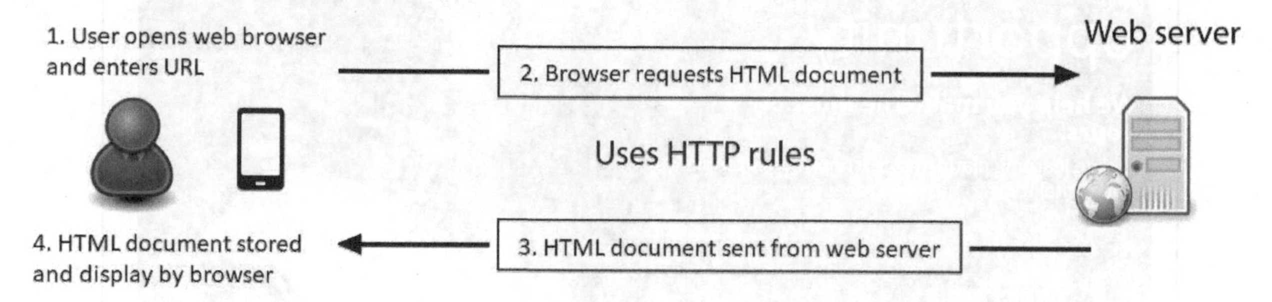

Email

Since developer Ray Tomlinson sent the first email message in 1971, email has become an essential part of everyday life. It is estimated that over 400 billion emails are sent daily. However, only 15 percent of this total (60 billion) are legitimate; the remaining 85 percent or 340 billion daily emails are spam.[2]

Note 3

Using email for political purposes helped introduce email to the general public. Former President Carter used a basic email system, which charged $4 per message, to coordinate strategies and send speeches during his 1976 presidential campaign.

The two basic components involved in sending and receiving email are the **Mail User Agent (MUA)** and **Mail Transfer Agent (MTA)**. An MUA is used to read and send mail from a device, such as an app (like Microsoft Outlook) or a webmail interface (like Gmail). MTAs are programs that accept email messages from senders and route them to their recipients.

Originally, email was sent to a mail server using the *Simple Mail Transfer Protocol* (*SMTP*). When it arrived at its destination, it was downloaded from the server using the *Post Office Protocol* (*POP3*). MUAs needed to use both SMTP and POP3 so that they could send user messages to the user's mail server (using SMTP) and could download messages intended for the user from the user's mail server (using POP3).

Later, **Internet Mail Access Protocol (IMAP)** was introduced and essentially replaced POP3. IMAP allows users to leave email on the mail server so that they can read email from multiple devices. (They could read an email on a smartphone when it was first received and then later reference the email on a laptop.) IMAP also allowed messages to be organized into folders that could again be accessed consistently from any endpoint. More recently, webmail has become more popular and widespread. Users can use a website as their MUA (such as Gmail) and no longer must configure devices with SMTP or IMAP server settings.

As email is transferred from one MTA to another MTA (such as from the sender's device to the recipient's device), information is added to the email header. The email header contains information about the sender, recipient, email's route through MTAs, and various authentication details. Figure 4-3 shows a partial email header.

Figure 4-3 Email header

```
Received: from BYAPR15MB3462.namprd15.prod.outlook.com (2603:10b6:a03:112::10)
by BN7PR15MB4081.namprd15.prod.outlook.com with HTTPS; Fri, 11 Dec 2020
12:42:27 +0000
Received: from BN6PR13CA0038.namprd13.prod.outlook.com (2603:10b6:404:13e::24)
by BYAPR15MB3462.namprd15.prod.outlook.com (2603:10b6:a03:112::10) with
Microsoft SMTP Server (version=TLS1_2,
cipher=TLS_ECDHE_RSA_WITH_AES_256_GCM_SHA384) id 15.20.3632.18; Fri, 11 Dec
2020 12:42:26 +0000
Received: from BN7NAM10FT025.eop-nam10.prod.protection.outlook.com
(2603:10b6:404:13e:cafe::a9) by BN6PR13CA0038.outlook.office365.com
(2603:10b6:404:13e::24) with Microsoft SMTP Server (version=TLS1_2,
cipher=TLS_ECDHE_RSA_WITH_AES_256_GCM_SHA384) id 15.20.3654.9 via Frontend
Transport; Fri, 11 Dec 2020 12:42:25 +0000
Authentication-Results: spf=softfail (sender IP is 161.6.94.39)
smtp.mailfrom=potomac1050.mktomail.com;
topperwkuedu94069.mail.onmicrosoft.com; dkim=fail (body hash did not verify)
header.d=raritan.com;topperwkuedu94069.mail.onmicrosoft.com; dmarc=none
action=none header.from=raritan.com;
Received-SPF: SoftFail (protection.outlook.com: domain of transitioning
potomac1050.mktomail.com discourages use of 161.6.94.39 as permitted sender)
Received: from email.wku.edu (161.6.94.39) by
BN7NAM10FT025.mail.protection.outlook.com (10.13.156.100) with Microsoft SMTP
Server (version=TLS1_2, cipher=TLS_ECDHE_RSA_WITH_AES_256_CBC_SHA384) id
15.20.3654.12 via Frontend Transport; Fri, 11 Dec 2020 12:42:25 +0000
Received: from e16-12.ad.wku.edu (161.6.94.62) by e16-05.ad.wku.edu
(161.6.94.39) with Microsoft SMTP Server (version=TLS1_2,
cipher=TLS_ECDHE_RSA_WITH_AES_256_CBC_SHA384_P256) id 15.1.1979.3; Fri, 11
```

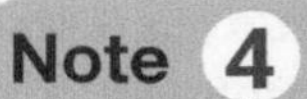

Note 4

Each MTA along the path of an email adds its own information to the top of the email header. This means that when reading an email header, the final destination MTA information will be at the top of the header, and reading down through each subsequent header will ultimately end at the sender MTA header information.

Email headers also contain an analysis of the email by the MTA. Table 4-1 shows the different analysis categories and abbreviations used by the Microsoft Outlook 365 MUA.

Table 4-1 Microsoft Office 365 email analysis

Abbreviation	Category
BULK	Bulk
DIMP	Domain impersonation
GIMP	Mailbox intelligence-based impersonation
HPHISH	High confidence phishing
HSPM	High confidence spam
MALW	Malware
PHSH	Phishing
SPM	Spam
SPOOF	Spoofing
UIMP	User impersonation
AMP	Anti-malware
SAP	Safe attachments
OSPM	Outbound spam

Email **attachments** are documents added to an email message, such as word processing documents, spreadsheets, or pictures. These attachments are encoded in a special format and sent as a single transmission along with the email message itself. When the receiving computer receives the attachment, it converts it back to its original format.

Internet Security Risks

Users face several risks when using the Internet. These can be classified into user device threats, threats from web servers, and transmission risks.

User Device Threats

Multiple threats focus on the user device itself. The two most prominent are browser dangers and email risks.

Browser Dangers

In the webpage transmission process, the device (computer, smartphone, tablet, etc.) does not view the HTML document on the web server; rather, the entire document is transferred and then stored on the device before the web browser displays it. This transfer-and-store process creates opportunities for attackers to send malicious code to the user's device, making web browsing a potentially dangerous experience.

The danger associated with using a web browser has been magnified by functionality added to the web experience. In the early days of the web, users viewed *static* content (information that does not change) such as text and pictures through a web browser. In short, every user saw the same content. As the web increased in popularity, the demand rose for *dynamic* content that can change. Dynamic websites deliver and display the content of each page on the fly (dynamically) according to user behavior.

Note 5

Dynamic websites can display different information based on various parameters. One parameter is the current or past actions of the user on the site: if the user has been a previous visitor to the site, she may receive a *Welcome Back* message with coupons for items she was previously viewing. Another parameter is the location of the user: a dynamic multilingual website can display webpages in different languages depending on where the user is located. Dynamic websites can also display different content depending on whether the device accessing the site is a handheld smartphone or a full-size laptop computer.

The solution to producing dynamic content came in several forms. One solution was to allow scripting code to be downloaded from the web server into the user's web browser. Another solution took the form of "additions" known as extensions that could be added to a web browser to support dynamic content. However, each of these solutions carries cybersecurity risks.

Scripting Code One means of adding dynamic content is for the web server to download a "script" or series of instructions in the form of computer code that commands the browser to perform specific actions. Two popular scripting languages are JavaScript and Windows PowerShell.

JavaScript The most popular scripting language for generating code is **JavaScript**. Because JavaScript cannot create separate "stand-alone" applications, the JavaScript instructions are embedded inside HTML documents. These interact with the HTML page's *Document Object Model (DOM)*, which connects webpages to scripts or programming languages. When a website that uses JavaScript is accessed, the HTML document that contains the JavaScript code is downloaded onto the user's computer.

Visiting a website that automatically downloads code to run on the user's device can be dangerous: an attacker could write a malicious script and have it downloaded and executed on the user's computer. A malicious JavaScript script could capture and remotely transmit user information without the user's knowledge or authorization. For example, an attacker could capture and send the user's email address to a remote source or even send a fraudulent email from the user's email account. Other JavaScript attacks can be even more malicious. An attacker's JavaScript program could scan the user's network and then send specific commands to disable security settings or redirect a user's browser to an attacker's malicious website.

Note 6

Different defense mechanisms are intended to prevent JavaScript programs from causing serious harm. JavaScript does not support certain capabilities. For example, JavaScript running on a local computer cannot read, write, create, delete, or list the files on that computer. Also, JavaScript can be run in a restricted environment ("sandbox") to limit what computer resources it can access or actions it can take. However, there are still security concerns with JavaScript.

PowerShell A task automation and configuration management framework from Microsoft is known as **PowerShell**. Administrative tasks in PowerShell are performed by *cmdlets* ("command-lets"), which implement a specific operation. Users and developers can create and add their own cmdlets to PowerShell, and the PowerShell runtime can even be embedded inside other applications. On the Microsoft Windows platform, PowerShell has full access to a range of operating system operations and components.

Note 7

PowerShell is an open-source and cross-platform product running on Windows, macOS, and Linux platforms.

The power and reach of PowerShell make it a prime target for threat actors. PowerShell can be configured so that its commands are not detected by any antimalware running on the computer. Because most applications flag PowerShell as a "trusted" application, its actions are rarely scrutinized.

Note 8

One recent attack illustrates how PowerShell can be used by threat actors. The script ran with the PowerShell parameters *ExecutionPolicyByPass* (*allow the PowerShell script to run despite any system restrictions*), *WindowStyleHidden* (*run the script quietly without any notification to the user*), and *NoProfile* (*do not load the system's custom PowerShell environment*).

Extensions **Extensions** expand the normal capabilities of a web browser. A browser extension is essentially a small piece of software code that performs a function or adds a feature to a browser to support dynamic actions. Extensions are browser-dependent: an extension that works in the Google Chrome web browser does not function in the Microsoft Edge browser.

Because extensions act as part of the browser itself, they generally have wider access privileges than JavaScript running in a webpage. Since extensions are given special authorizations within the browser, they are also attractive tools used by attackers.

Note 9

At one time, *plug-ins* were widely popular as additions to web browsers. A plug-in adds new functionality to the web browser so that users can play music, view videos, or display special graphical images within the browser that it normally could not play or display. Technically, a plug-in is a *third-party binary library* that lives outside of the "space" that a browser uses on the computer for processing and serves as the link to external programs that are independent of the browser. A single plug-in can be used on different web browsers. However, almost all major web browsers have dropped support for automatic plug-ins due to their security risks.

Email Risks

Another common means of distributing attacks is through email. Email risks include malicious attachments, embedded hyperlinks, and spam.

Malicious Attachments Attacks are often distributed through email attachments. Opening an attachment that contains malware can immediately infect the computer with that malware.

One common means of distributing malware through malicious attachments is through Microsoft Office files that contain a **macro**. A macro is a series of instructions that can be grouped as a single command, much like a script. Macros are used to automate a complex or repeated series of tasks. They are stored within the user document (such as in an Excel .xlsx worksheet or Word .docx file) and can be launched automatically when the document is opened.

Macros are usually written by using *Visual Basic for Applications* (*VBA*), which is an "event-driven" Microsoft programming language. VBA is built into most Microsoft Office applications (including Word, Excel, and PowerPoint) for both Windows and Apple macOS platforms. It can implement a wide variety of tasks, including manipulating user interface features such as toolbars, menus, forms, and dialog boxes. VBA can even control one application from another application: VBA can automatically create a Microsoft Word report from data in a Microsoft Excel spreadsheet, for example.

Caution !

Although macros date back to the late 1990s, they continue to be a key attack vector. Microsoft has reported that 98 percent of all Office-targeted threats are a result of macro-based malware, and it has warned users that Office macros particularly in Excel are commonly used to compromise Windows systems.[3]

Most users are unaware of the danger of attachments and routinely open any email attachment that they receive, even if it is from an unknown sender. Attackers often include in the subject line information that entices even reluctant users to open the attachment, such as a current event (*Check out these images from yesterday's fire!*) or information about the recipient (*Is that really you in this picture?*).

Caution !

Some email-distributed malware replicates by sending itself as an email attachment to all of the contacts in a user's email address book. The unsuspecting recipients, seeing that an email and attachment was sent from a friend's email account, assume it is trustworthy and open the malicious attachment that then infects their computer.

Embedded Hyperlinks Many email messages have **embedded hyperlinks**, which are contained within the body of the message as a shortcut to a website. However, attackers can take advantage of embedded hyperlinks to direct users to a malicious site. This site can impersonate the actual site to trick users into entering a password or private information that the attacker steals. Or the site could simply push malware on the user's computer.

Note 10

Despite the risks of hyperlinks, deciding not to use them can be a difficult habit to break. One organization distributed an email from the IT security department that specifically warned users not to click embedded hyperlinks because of the danger associated with them. However, at the bottom of the email, it said, "For more information, click on this link."!

Redirection from a malicious hyperlink is easily accomplished because an embedded hyperlink in an email message can display any content or URL to the user. However, what is displayed in the hyperlink is not the actual address of the website. Figure 4-4 shows a phishing email message with multiple embedded hyperlinks. Although the hyperlink is displayed as www.CapitalOneSettlement.com/claim, the underlying URL, which redirects the user when he clicks it, is completely different.

Figure 4-4 Embedded hyperlink

How To Get Benefits: You must submit a Claim Form, including any required documentation. (You do not need to file a Claim to access the Restoration Services.) The deadline to file a Claim Form is **August 22, 2022.** You can easily file a Claim online at www.CapitalOneSettlement.com/claim. You can also get a paper Claim Form at the website or by calling toll free 1-855-604-1811, and file by mail. **When filing your Claim use your Unique ID Number and your PIN (located at the top of this email).**

Your Other Options. If you file a Claim Form, object to the Settlement and attorneys' fees and expenses, or do nothing, you are choosing to stay in the Settlement Class. You will be legally bound by all orders of the Court and you will not be able to start, continue or be part of any other lawsuit against Capital One, Amazon, or related parties about the Data Breach. If you don't want to be legally bound by the Settlement or receive any benefits from it, you must exclude yourself by **July 7, 2022.** If you do not exclude yourself, you may object to the Settlement and attorneys' fees and expenses by **July 7, 2022**. The Court has scheduled a hearing in this case for **August 19, 2022**, to consider whether to approve the Settlement, attorneys' fees of up to 35% of the Settlement Fund plus costs and expenses, Service Awards of up to $5,000 for the Class Representatives and other class members deposed in the case, as well as any objections. You or your own lawyer, if you have one, may ask to appear and speak at the hearing at your own cost, but you do not have to.

How can I get More Information? For complete information about all of your rights and options, as well as Claim Forms, the Long Form Notice and Settlement Agreement, visit www.CapitalOneSettlement.com, or call 1-855-604-1811.

NOTICE AUTHORIZED BY: United States District Court for the Eastern District of Virginia

Caution !

Note also the wording contained in Figure 4-4's phishing email. First, it is designed to scare the user into making a quick response by giving several deadlines by which action must be taken. Also, it says that action must be taken to either include or exclude oneself from the claim (the only two options), implying that *some* response is necessary. And the email says that if the recipient does not exclude themselves, then *"you may be object* (sic, should be "subject") *to the Settlement and attorney's fees and expenses,"* which is never the case in a settlement.

Spam The amount of **spam**, or unsolicited email, that goes through the Internet, can be measured in the hundreds of billion messages sent *daily*. The reason users receive so many spam messages that advertise drugs, cheap mortgage rates, and items for sale is that sending spam is a lucrative business. It costs spammers very little to send millions of spam email messages. Almost all spam is sent from botnets: a spammer who does not own his own botnet can lease time from other attackers ($40 per hour) to use a botnet of up to 100,000 infected computers to launch a spam attack. Even if spammers receive only a small percentage of responses, they still make a large profit. For example, if a spammer sent spam to 6 million users for a product with a sale price of $50 that cost only $5 to make, and if only 0.001 percent of the recipients responded and bought the product (a typical response rate), the spammer would make over $270,000 in profit.

Text-based spam messages that include words such as *Viagra* or *investments* can easily be trapped by email **spam filters**. These filters typically look for spam-related words and block the email. Because of the increased use of these filters, spammers have turned to another approach for sending their spam. Known as *image spam*, it uses graphical images of text to circumvent text-based filters. Image spam cannot be filtered based on the content of the message because it appears as an image instead of text. These spam messages often include nonsense text so that it appears the email message is legitimate (an email with no text can prompt the spam filter to block it). Figure 4-5 shows an example of image spam.

Figure 4-5 Image spam

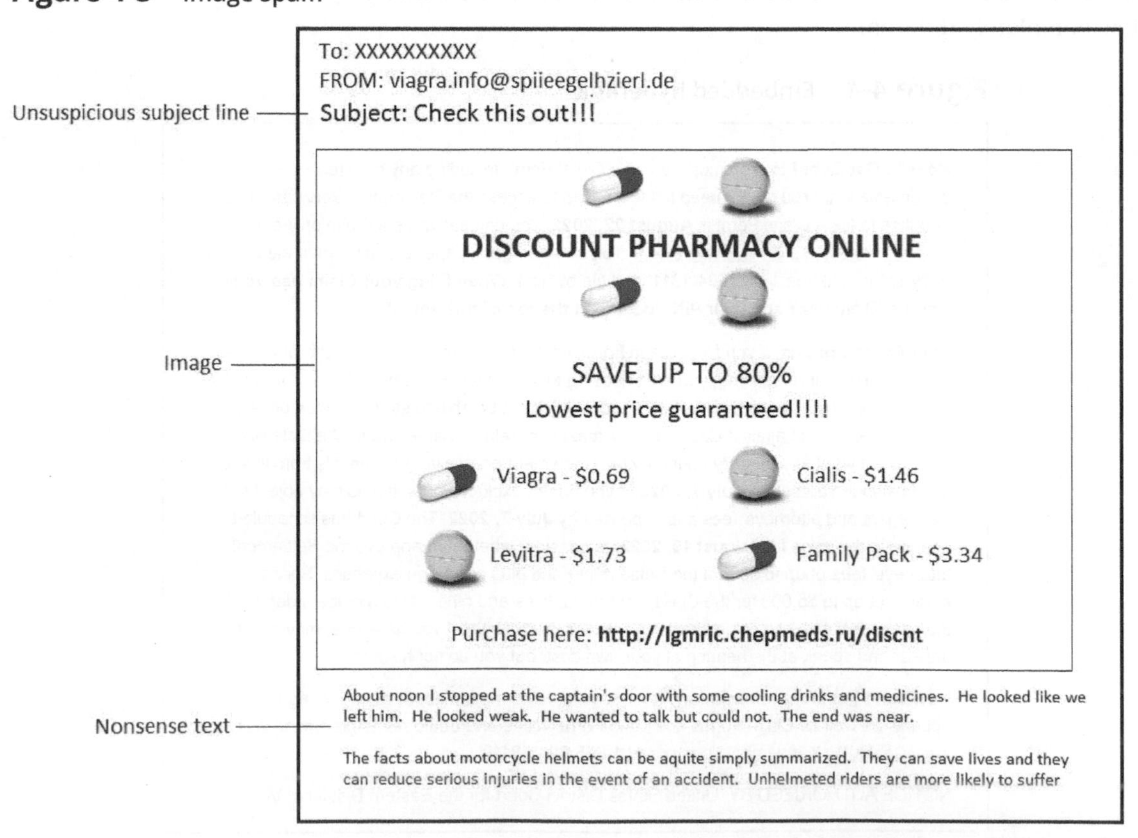

Beyond just being annoying, spam significantly reduces work productivity as users spend time deleting spam messages. Spam is also costly to organizations that must install and monitor technology to block spam. However, one of the greatest risks of spam is that it is used to widely distribute malware.

Threats from Web Servers

Web servers that provide content to users can also pose a risk. These threats from web servers include malvertising, drive-by downloads, cross-site scripting (XSS) attacks, and cross-site request forgery (CSRF) attacks.

Malvertising

When visiting a typical website, it is common to see multiple advertisements around the pages. For example, visiting a fitness tracking website often results in ads being displayed that promote athletic shoes, sports drinks, weight loss, and other related products.

These ads do not usually come from the main site itself; instead, most mainstream and high-trafficked websites outsource the ad content on their pages to several third-party advertising networks. When users go to the site's page, their web browsers silently connect to dozens of advertising network sites from which ad banners, pop-up ads, video files, and pictures are sent to the users' computers to display advertisements.

Attackers have turned to using these third-party advertising networks to distribute their own malware—such as ransomware, Trojans, and keyloggers—to unsuspecting users who are visiting a well-known website. Attackers may infect third-party advertising networks so that their malware is distributed through ads sent to users' devices. Or the attackers may promote themselves as reputable third-party advertisers while in reality, they are distributing their malware through the ads. This is known as **malvertising** (*malicious advertising*).

Note 11

The New York Times, Reuters, Bloomberg, and Google, among many others, have all been infected with malvertising. Because of its widespread use, it is difficult to determine the costs of malvertising, but it is estimated that the losses range from $6.5 billion to $19 billion. And at least one out of every 100 ads displayed contains malicious code.[4]

Malvertising has several advantages for the attacker, including the following:

- Malvertising occurs on "big-name" websites, most frequently news publications that attract many visitors each day who are keeping abreast of the latest news stories. These unsuspecting users, who otherwise would avoid or be suspicious of less popular sites, are deceived into thinking that because they are on a reputable site, they are free from attacks.
- Website owners do not know malware is being distributed on their website through ads. This is because they do not know what type of ad content a third-party ad network is displaying on their site at any given time.
- Ad networks rotate content quickly, so that not all visitors to a site are infected, making it difficult to determine if malvertising was actually the culprit of an attack. Even when an ad is pinpointed in an investigation as malicious, it is virtually impossible to prove which ad network was responsible.
- Because advertising networks configure ads to appear according to the user's computer (which browser or operating system they are using) or identifying attributes (their country locations or search keywords they used to find the site), attackers can narrowly target their victims. For example, an attacker who wants to target U.S. federal government employees might distribute ads with malicious content for anyone who entered "Government travel allowance" into a search engine.

Note 12

Because these attacks can precisely target their victims, often "high-value victims" are pinpointed. For example, an attacker may place malicious ads before individuals who are conducting a keyword search for hotel rates at an upcoming security conference.

Preventing malvertising is a difficult task. Website operators are unaware of the types of ads that are being displayed, users trust going to a "mainstream" website, and turning off ads in a web browser may interfere with the user's web experience.

Drive-By Downloads

Whereas malvertising seeks to infect a mainstream website through third-party advertising networks, other attacks attempt to infect the website directly. In some instances, this can result in a user's device becoming infected just by viewing the website. Such occurrences are called **drive-by downloads**.

Attackers first identify a well-known website and then attempt to inject malicious content by exploiting it through a vulnerability in the web server. These vulnerabilities often permit the attacker to gain direct access to the web server's underlying operating system and then inject malicious content into the compromised website. When unsuspecting users visit an infected website, their browsers download code usually written in JavaScript that targets a vulnerability in the user's browser. If the script can run successfully on the user's computer, it may instruct the browser to connect to the attacker's own web server to download malware, which is then automatically installed and executed on the user's computer.

Note 13

Unlike a traditional download that asks for the user's permission to perform an action, a drive-by download can be initiated simply by *visiting* an infected website. Users are not required to give permission for the download to occur.

Cross-Site Scripting (XSS)

Websites that create dynamic content typically ask for user input and then create the content based on that input. Figure 4-6 illustrates a web form that asks for the user's first name, last name, and email address. The user "Abby" enters her information and may then receive a message, "Thank you, Abby, for registering."

Figure 4-6 Web form

Register
First name
Last name
Email address
Register

However, if the website does not first validate the user's input (called *sanitizing*), attackers can exploit input that has been entered into a form. In the previous example, input the user enters for *Name* is automatically added to a code segment that becomes part of an automated response back to the user's web browser, which it executes. If the input was not a name but was malicious code instead, this code would be sent to the user's web browser and executed. This is called a **cross-site scripting (XSS)** attack.

Note 14

The term *cross-site scripting* refers to an attack using scripting that originates on one site (the web server) to impact another site (the user's computer).

Cross-Site Request Forgery (CSRF)

A **cross-site request forgery (CSRF)** (sometimes pronounced *sea-surf*) takes advantage of an authentication "token" that a website sends to a user's web browser. If a user has logged into a website using a valid username and password but is then tricked into loading another webpage, the new page inherits the identity and privileges of the victim who logged in. The victim could then perform an undesired function on the new webpage on the attacker's behalf. Figure 4-7 illustrates a cross-site request forgery.

Figure 4-7 Cross-site request forgery (CSRF)

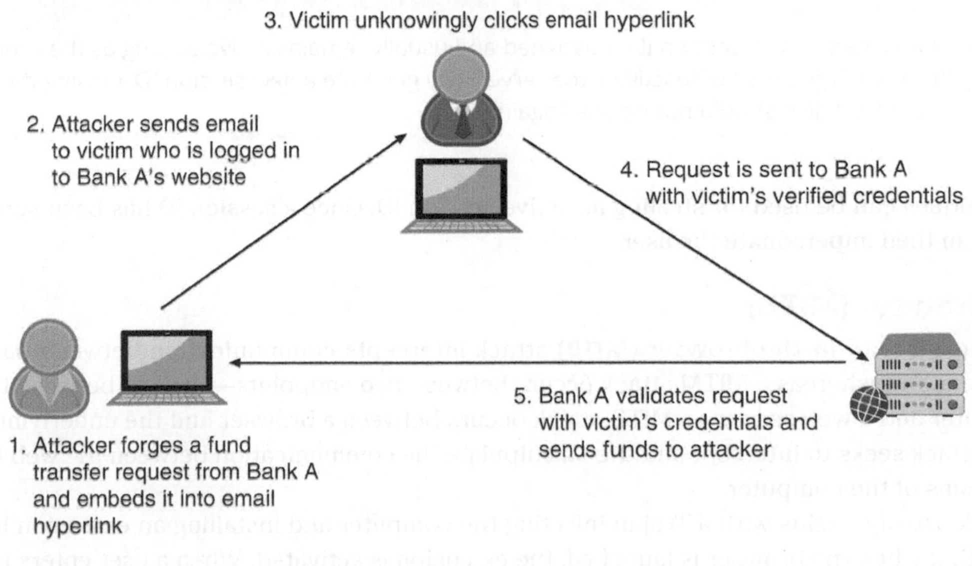

Note 15

In a CSRF attack, a *request* to a website is not from the authentic user but is a *forgery* that involves *crossing sites*.

Transmission Risks

Some attacks are designed to intercept network communications across the Internet. These transmission attacks include man-in-the-middle, session replay, and man-in-the-browser attacks.

Man-in-the-Middle (MITM)

Suppose that Angie, a high school student, is in danger of receiving a poor grade in math. Her teacher, Mr. Ferguson, mails a letter to Angie's parents requesting a conference regarding her performance. However, Angie waits for the mail and retrieves the letter from the mailbox before her parents come home. She forges her parent's signature on the original letter declining a conference and mails it back to her teacher. Angie then replaces the real letter with a counterfeit pretending to be from Mr. Ferguson that compliments Angie on her math work. The parents read the fake letter and tell Angie they are proud of her, while Mr. Ferguson is puzzled why Angie's parents are not concerned about her grades.

Angie has conducted a type of **man-in-the-middle (MITM)** attack. In a MITM, a threat actor will position herself in a communication pathway between two parties. Neither of the legitimate parties is aware of the threat actor and thus communicates freely, thinking they are talking only to the authentic party. The goal of a MITM attack is to either eavesdrop on the conversation or impersonate one of the parties.

Session Replay

A *replay* attack is a variation of a MITM attack. Whereas a MITM attack alters and then sends the transmission immediately, a replay attack makes a copy of the legitimate transmission before sending it to the recipient. This copy is then used later (when the MITM "replays" the transmission).

A specific type of replay attack is a **session replay** attack, which involves intercepting and using a *session ID* to impersonate a user. A session ID is a unique number that a web server assigns a specific user for the duration of that user's visit (session). Most servers create complex session IDs by using the date, time of the visit, and other variables such as the device IP address, email address, username, user ID, role, privilege level, access rights, language preferences, account ID, current state, last login, session timeouts, and other internal session details. A sample session ID is *fa2e76d49a0475910504cb3ab7a1f626d174d2d*.

Note 16

Each time a website is visited, a new session ID is assigned and usually remains active as long as the browser is open. In some instances, after several minutes of inactivity, the server may generate a new session ID. Closing the browser terminates the active session ID, which should not be used again.

Several techniques can be used for stealing an active session ID. Once a session ID has been successfully stolen, the threat actor can then impersonate the user.

Man-in-the-Browser (MITB)

Like a MITM attack, a **man-in-the-browser (MITB)** attack intercepts communication between parties to steal or manipulate the data. But whereas a MITM attack occurs between two endpoints—such as between two user laptops or a user's computer and a web server—a MITB attack occurs between a browser and the underlying computer. Specifically, a MITB attack seeks to intercept and then manipulate the communication between the web browser and the security mechanisms of the computer.

A MITB attack usually begins with a Trojan infecting the computer and installing an extension into the browser configuration so that when the browser is launched, the extension is activated. When a user enters the URL of a site, the extension checks to determine if this is a site targeted for attack. After the user logs in to the site, the extension waits for a specific webpage to be displayed in which a user enters information, such as the account number and password for an online financial institution (a favorite target of MITB attacks). When the user clicks "Submit," the extension captures all the data from the fields on the form and may even modify some of the entered data. The browser then sends the data to the server, which performs the transaction and generates a receipt that is sent back to the browser. The malicious extension again captures the receipt data and modifies it (with the data the user originally entered) so that it appears that a legitimate transaction has occurred.

There are several advantages to a MITB attack:

- Most MITB attacks are distributed through Trojan browser extensions, which provide a valid function to the user but also install the MITB malware, making it difficult to recognize that malicious code has been installed.
- Because MITB malware is selective as to which websites are targeted, an infected MITB browser might remain dormant for months until triggered by the user visiting a targeted site.
- MITB software resides exclusively within the web browser, making it difficult for standard anti-malware software to detect it.

Two Rights & A Wrong

1. When a website that uses JavaScript is accessed, the HTML document that contains the JavaScript code is downloaded onto the user's computer.
2. Extensions expand the normal capabilities of a web browser.
3. Spam, while annoying and a drain on productivity, is not considered dangerous.

See Appendix A for the answer.

Internet Defenses

Because Internet-based attacks are the most common types of attacks that users face, multiple strong layers of defense must be implemented. However, defending against Internet-based attacks begins with the foundation of first having the device itself properly secured. This includes managing patches, running antimalware software, examining personal firewall settings, and having data backups.

Note 17

Basic computer security is covered in detail in Module 3.

Once the device is properly secured, then the additional defensive security steps to resist Internet-based attacks fall into two broad categories. These defenses include securing the web browser and creating email defenses.

Note 18

Not all Internet threats can be counteracted by users. For example, threats from web servers (malvertising, drive-by downloads, XSS, and CSRF) along with transmission risks (MITM, session replay, and MITB) lie beyond the scope of a user fixing a vulnerability on a web server or preventing a session replay attack. But what a user can do is to keep her device up to date with the latest patches and be cautious.

Securing the Web Browser

Modern web browsers have evolved into strong defenses against attacks. This protection can be divided into web browser security-related indicators, security settings, and managing browser extensions.

Security-Related Indicators

The dashboard of a modern car is filled with different indicators. These include indicators for low tire pressure, high engine temperature, low oil pressure, malfunctioning airbags, and several others. These indicators tell the driver that something is not normal. However, these indicators are of little value if the driver does not know what they mean. A low tire pressure indicator that is displayed is often ignored by a driver who is unaware of what it means, thinking (or perhaps hoping!) that it will go away soon. Knowing the indicators on a car dashboard can head off more serious problems.

The same is true with a web browser. This software displays indicators to the user about something that may need attention or is a warning about a danger. Like driving a car, it is important to know what these indicators are telling users. Web browsers display several security-related indicators to users regarding the safety or risk of the website they are viewing. These indicators include Hypertext Transport Protocol Secure (HTTPS) padlocks and web browser warnings.

Hypertext Transport Protocol Secure (HTTPS) Padlocks Originally, web servers distributed HTML documents based on the HTTP protocol. However, information sent via HTTP was unprotected and could be available for someone to capture and view. If the transmission involved SSNs or credit card numbers, the protocol could prove to be a serious risk.

However, today most web transmissions use a more secure communications method between a browser and a web server than original HTTP. This improved version is "plain" HTTP sent using a secure cryptographic protocol and is called **Hypertext Transport Protocol Secure (HTTPS)**. HTTPS URLs begin with *https://* instead of *http://*.

Note 19

Another secure protocol for HTTP was Secure Hypertext Transport Protocol (SHTTP). However, it was not as secure as HTTPS and is now obsolete.

At one time, web browsers prominently displayed a visual indicator to alert users that the connection between the browser and the web server was using HTTPS. As shown in Figure 4-8, most browsers displayed a green padlock to indicate the connection was secure.

Figure 4-8 HTTPS padlock

However, as more websites transitioned to HTTPS, some web browsers changed from displaying an indicator that the connection *was* secure to only a warning that the connection *was not* secure, as seen in Figure 4-9. The Chrome browser, like Microsoft Edge and other browsers, still displays a padlock though it is gray rather than green.

Figure 4-9 Google Chrome HTTP warning

Note 20

The change away from the green padlock is much like the indicator lights on a car dashboard: car indicator lights are designed to turn on only when to signal a problem. This approach of keeping the lights off when "all is OK" but only turn on the warning light when something demands attention is the same reason for migrating away from web browser indicators when there is no problem.

Web Browser Warnings Web browsers also display warnings if a website is known or suspected to be dangerous. Browsers can do this because they maintain a list of suspicious websites that is updated constantly. In most instances, the screen turns bright red with the warning prominently displayed. Table 4-2 lists the warnings and explanations for the Google Chrome web browser.

When you receive a warning message in a web browser, it is critical to carefully read the message and understand what it says. Although it is possible to bypass the warning and view an unsafe site or download a suspicious file, *it is strongly recommended that these warnings be heeded.*

Web Browser Security Settings

Modern web browsers have evolved into customizable applications that allow the user to tailor settings based on personal preferences. Beyond basic settings such as the preferred home page and size of displayed characters, browsers also allow the user to customize cybersecurity settings.

Table 4-2 Google Chrome web browser warnings

Warning message	Explanation
The site ahead contains malware	The site you start to visit might try to install malware on your computer.
Deceptive site ahead	The site you are trying to visit might be a phishing site.
Suspicious site	The site you want to visit seems suspicious and may not be safe.
The site ahead contains harmful programs	The site you start to visit might try to trick you into installing programs that cause problems when you're browsing online.
This page is trying to load scripts from unauthenticated sources	The site you are trying to visit isn't secure.
Fake site ahead (or may display *Did you mean [site]* or *Is this the right site?*)	"Appears similar to a safe site you usually visit" or "Tries to trick you with a URL that is slightly changed from a known safe site" or "Has a URL that is slightly different from a URL in your browsing history."

However, the number of options for correctly customizing a secure web browser can quickly become overwhelming and sometimes even confusing. Modern browsers now implement *modes* of cybersecurity that encompass multiple settings. This only requires a user to select an appropriate mode and not attempt to configure multiple individual settings. Figure 4-10 shows the modes for the Microsoft Edge web browser, *Balanced* and *Strict*. If *Strict* mode is chosen, an indicator *Added security* also appears in the browser navigation bar, as shown in Figure 4-11.

Figure 4-10 Microsoft Edge security modes

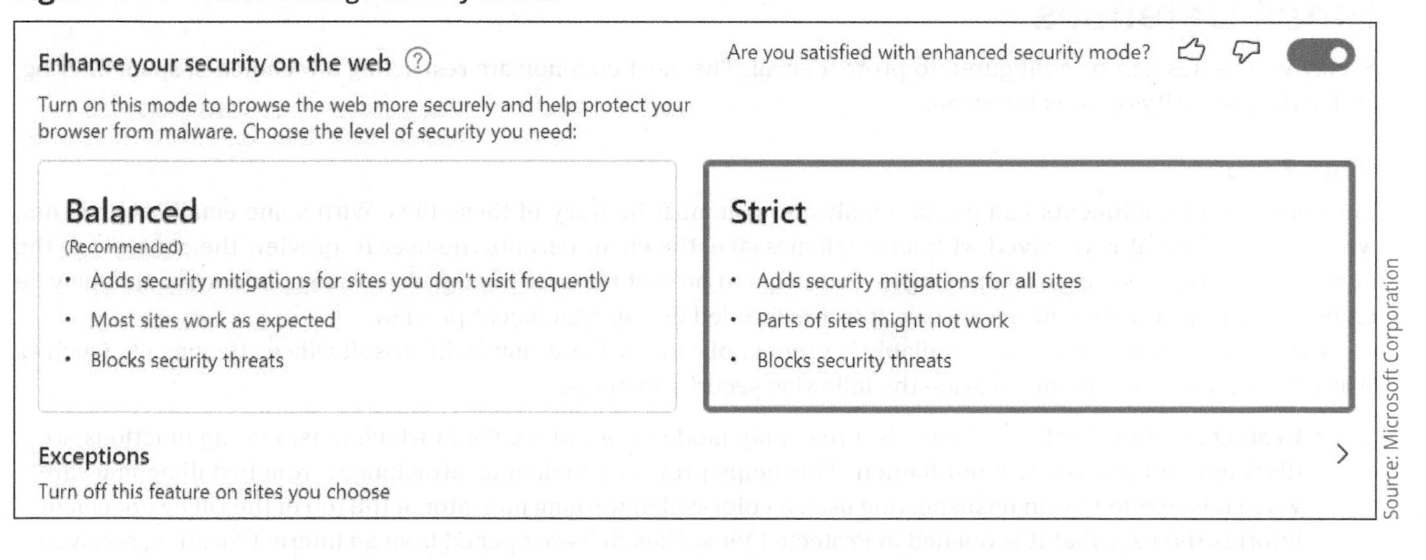

Source: Microsoft Corporation

Figure 4-11 Microsoft Edge Strict mode indicator

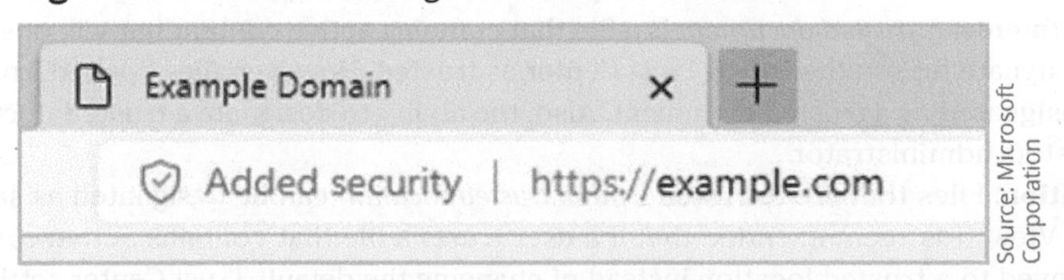

Source: Microsoft Corporation

Caution !

Microsoft Edge requires that the enhanced security mode be first turned on, and then the choice of *Balanced* or *Strict* be made. In contrast, Google Chrome has three modes: *Enhanced protection*, *Standard protection*, and *No protection*. These two browsers offer three options (essentially off, low, and high) but are presented in a slightly different format.

It is recommended that the highest level of security mode be turned on in a web browser. If this proves to be too restrictive, then exceptions can be made to this highest level. For example, in Microsoft Edge, an exception list can be made for accessing certain websites that are known to be trustworthy. If creating exceptions still affects the browsing experience, then the lower-level mode can be implemented. However, *having the security mode turned off completely is not recommended.*

Managing Browser Extensions

Browser extensions carry security risks since they are often from third parties other than the web browser vendor (such as Microsoft or Google). The following are ways to minimize the risks of browser extensions:

- **Check the browser first**. Before installing a new extension, first, check to see if this feature has already been added to the browser itself. Browser vendors are continually adding new features that once required extensions, such as managing to-do lists or saving news articles for later reading.
- **Avoid using too many extensions**. It is a good practice to keep the number of installed extensions limited to only those that are needed.
- **Use reputable sources**. It is important to only install extensions that are from reputable sources such as the Google Chrome Web Store or Microsoft Store.
- **Review and purge unused extensions**. Periodically review the extensions that are installed and determine if they are used frequently enough to warrant their installation. If not, delete them.

Email Defenses

Security defenses can be configured to protect email. The most common are restricting attachments, spam filtering, and setting security options for email.

Attachments

Because email attachments can contain malware, you must be wary of these files. With some email MAU clients, when an attachment is received with an email message, the client permits the user to preview the contents of the attachment without saving and opening it. This helps to protect the user from malicious code or macros that may be embedded in the attachment because scripts are disabled during attachment preview.

Attachment protection is also available in other applications. For example, Microsoft Office attachments (such as Word, Excel, and PowerPoint) provide the following security features:

- **Protected View**. **Protected View** is a read-only mode for an Office file in which most editing functions are disabled and macros will not launch. This helps prevent a malicious attachment from installing malware when it is opened by an unsuspecting user. A color-coded warning indicator at the top of the Office document informs the user that it is opened in Protected View. Files that are opened from an Internet location, received as an email attachment, opened from a potentially unsafe location, opened from another user's OneDrive storage, or have "active content" (macros or data connections) display the Protected View warning message. Users can click the *Enable Editing* button to open the file for editing.
- **Trusted documents**. A *trusted document* is a file that contains active content but will open without a warning. Users can designate files in the Office Trust Center as trusted. However, files opened from an unsafe location cannot be designated as a trusted document. Also, the ability to designate a trusted document can be turned off by the system administrator.
- **Trusted location**. Files that are retrieved from a *trusted location* can be designated as safe and will not open in Protected View. It is recommended that if a user trusts a file that contains active content like a macro, it should be moved to a trusted location instead of changing the default Trust Center settings to allow macros.

Caution 

Unless there is a business requirement for macros, support for their use should be disabled across the Microsoft Office suite.

Spam Filters

Beyond being annoying and disruptive, spam can also pose a serious security risk: spammers can distribute malware as attachments through their spam email messages. Spam filtering applications can be implemented on both the user's device as well as at the corporate or the Internet service provider level.

Note 21

Most users actually receive only a small amount of spam in their local email inbox. The majority is blocked before it even reaches the user.

Email clients can be configured to filter spam that has bypassed a spam filter. The email client spam filtering settings often include the following features to block spam:

- **Blocked senders**. A list of senders can be entered for which the user does not want to receive any email. A message received from a blocked sender is instead sent to the junk email folder. Several databases are available on the Internet that include known spammers and others who distribute malicious content, and some sites allow users to download these lists and automatically add them to their email.
- **Allowed senders**. The opposite of a list of blocked senders is a list of only approved senders. A list of senders can be entered for which the user will accept email.
- **Blocked top-level domain list**. Email from entire countries or regions can also be blocked.

Note 22

Microsoft Outlook automatically blocks over 100 types of file attachments that may contain malware.

Email Security Settings

In addition to spam filters on the local email client, other security settings can be configured through an installed email MAU client on the device and through using web email.

Installed Email MAU Client When using an installed email MAU client on a device, the following settings can improve security:

- **Read messages using a reading pane**. Most email clients contain a *reading pane*, which allows the user to read an email message without actually opening it. Received email messages can be viewed safely in the reading pane because malicious scripts and attachments are not activated or opened automatically in the reading pane.

Note 23

Although malicious attachments may be blocked by using the email reading pane, messages and attachments from unknown or unsolicited senders should always be treated with caution.

- **Block external content**. Email clients can be configured to block external content in HTML email messages that are received, such as hyperlinks to pictures or sounds. When a user opens an email message or it is displayed in the reading pane, the computer downloads the external content so that the picture can be displayed or the sound played. In addition, spammers often send spam to a wide range of email addresses, not knowing which email addresses exist or are accurate. To determine which email addresses are valid and actually exist, a spammer can note which email accounts downloaded the external content and then add those email accounts to their spam list. Blocking external content helps to prevent this.

Web Email When accessing webmail through a web browser, the following should be considered:

- **Check account for unusual activity**. A good security practice is to periodically check the email account for any unusual activity. Were unfamiliar email messages sent from this account? Are there deleted email messages in the Trash folder? Was the account accessed from a different location or at a different time than normal? If any of these occur, the email password should be reset immediately.
- **Verify general settings**. The general email settings should also be regularly reviewed. Users should verify that any signature lines, accounts that receive forwarded email, contact list addresses, and other information are valid. If any suspicious information appears, the email password should be reset immediately.

Two Rights & A Wrong

1. Defending against Internet-based attacks begins with the foundation of first having the device itself properly secured.
2. HTTP is a secure protocol for sending information through the web.
3. Before installing a new extension, users should first check to see if this feature has already been added to the browser itself.

See Appendix A for the answer.

Module Summary

- The Internet is an international network of computer networks. These networks are operated by industry, governments, schools, and even individuals who all loosely cooperate. The World Wide Web (WWW) is composed of Internet server computers on networks that provide online information in the Hypertext Markup Language (HTML) format. Instructions written in HTML code specify how a local computer's web browser should display the various elements of a webpage. Web servers distribute HTML documents based on a set of standards known as the Hypertext Transport Protocol (HTTP).
- The two basic components involved in sending and receiving email are the Mail User Agent (MUA) and Mail Transfer Agent (MTA). An MUA is what is used to read and send mail from a device, such as an app or a webmail interface. MTAs are programs that accept email messages from senders and route them to their recipients. Originally email systems used the Simple Mail Transfer Protocol (SMTP) to handle outgoing mail and the Post Office Protocol (POP3) for incoming mail; today the Internet Mail Access Protocol (IMAP) is more commonly used instead.
- Users face several risks when using the Internet. In the webpage transmission process, the user's device does not view the HTML document on the web server; instead, the entire document is transferred and then stored on the device before the web browser displays it. This transfer-and-store process creates opportunities for attackers to send malicious code to the user's device, making web browsing a potentially dangerous experience. The danger associated with using a web browser has been magnified by functionality added to support dynamic content that can change.
- One means of adding dynamic content is for the web server to download a "script" or series of instructions in the form of computer code that commands the browser to perform specific actions. Two popular scripting languages are JavaScript and Windows PowerShell. Attackers could use both of these to inject code through a web browser onto a user's device. Another means to add dynamic content is through a browser extension, which is a small piece of software code that performs a function or adds a feature to a browser. Since extensions are given special authorizations within the browser, they are attractive attacking tools.

- Another common means of distributing attacks is through email. Attacks are often distributed through email attachments. Opening an attachment that contains malware can immediately infect the computer with that malware. One common means of distributing malware through malicious attachments is through Microsoft Office files that contain a macro. A macro is a series of instructions that can be grouped as a single command. Many email messages have embedded hyperlinks, which are contained within the body of the message as a shortcut to a website. Attackers can take advantage of embedded hyperlinks to direct users to a malicious site. Spam, or unsolicited email, is also used to distribute malware.
- Web servers that provide content to users can also pose a risk. One risk is malvertising. Attackers use third-party advertising networks to distribute malware through ads sent to users' web browsers. Other attacks attempt to infect the website directly. In some instances, this can result in a user's device becoming infected just by viewing the website; these attacks are called drive-by downloads. A cross-site scripting (XSS) attack occurs when a website does not validate a user's input into a form, allowing attackers to enter a code segment that is then executed by the user's web browser. A cross-site request forgery (CSRF) takes advantage of an authentication token that a website sends to a user's web browser.
- Some attacks are designed to intercept network communications across the Internet. A man-in-the-middle (MITM) attack occurs when a threat actor is positioned in a communication pathway between two parties to eavesdrop on the conversation or impersonate one of the parties. A session replay attack intercepts and uses a session ID to impersonate a user. A man-in-the-browser (MITB) attack intercepts communication between a browser and the underlying computer.
- Different defenses can be used to protect against Internet attacks. One defense is to understand and heed any web browser security-related indicators that provide important information. Hypertext Transport Protocol Secure (HTTPS) sends information across the web in a protected format. Most browsers display a gray padlock in a web browser when HTTPS is being used. Web browsers also display warnings if a website is known or suspected to be dangerous. There are many options for correctly customizing a secure web browser. Modern browsers now implement modes of cybersecurity that encompass multiple settings. Browser extensions also carry security risks, and these should be used only when necessary and deleted when no longer necessary.
- Security defenses can be configured to protect email. Because email attachments can contain malware, it is important to be wary regarding these files, even if they appear to come from a trusted source. Attachment protection is available in other applications like Microsoft Office and through email to MAU clients. Spam filtering applications can be implemented on both the user's device as well as at the corporate or the Internet service provider level. Security settings can also be configured through an installed email MAU client on the device, in addition to settings when accessing webmail through a web browser.

Key Terms

attachment
cross-site request forgery (CSRF)
cross-site scripting (XSS)
drive-by download
embedded hyperlink
extension
hyperlink
Hypertext Markup Language (HTML)
Hypertext Transfer Protocol (HTTP)
Hypertext Transport Protocol Secure (HTTPS)
Internet
Internet Mail Access Protocol (IMAP)
JavaScript
macro
Mail Transfer Agent (MTA)
Mail User Agent (MUA)
malvertising
man-in-the-browser (MITB)
man-in-the-middle (MITM)
PowerShell
Protected View
session replay
spam filter
spam
Transmission Control Protocol/Internet Protocol (TCP/IP)
uniform resource locator (URL)
web browser
World Wide Web (WWW)

Review Questions

1. Which of the following is NOT true about the Internet?
   a. It is not controlled by a single organization or government entity.
   b. Industry, governments, schools, and individuals all loosely cooperate in Internet self-governance.
   c. It is composed of networks to which devices are attached
   d. It is a local network of computers and networks.

2. What is the format of the information on World Wide Web pages?
   a. Hypertext Transfer Protocol (HTTP)
   b. Transmission Control Protocol/Internet Protocol (TCP/IP)
   c. Microsoft Web Manager (MWM)
   d. Hypertext Markup Language (HTML)

3. Which of the following is the more recent and advanced electronic email system?
   a. Simple Mail Transfer Protocol (SMTP)
   b. Internet Mail Access Protocol (IMAP)
   c. Post Office Protocol (POP)
   d. Transmission Control Protocol (TCP)

4. Which is the most popular scripting code used with webpages but also can be exploited?
   a. Microsoft Text Filer (MTF)
   b. JavaScript
   c. PowerShell
   d. MONTE

5. Web servers distribute HTML documents based on what set of standards?
   a. Hypertext Markup Language (HTML)
   b. Hypertext Transport Protocol (HTTP)
   c. Hypertext Distribution System (HDS)
   d. Hypertext Control Protocol/Internet Protocol (HCP/IP)

6. Which of the following statements about webpages is NOT true?
   a. A URL is the web address of a webpage.
   b. Web browsers send requests to remote web servers using HTTP.
   c. The user's web browser, using the code contained in the HTML document, displays the results on the web browser.
   d. A web browser displays the HTML document through a window that views the document on the web server.

7. Delysia has been asked to install Microsoft Outlook on a user's new computer so that he can read his email. What is Delysia installing?
   a. MTA
   b. ATA
   c. MUA
   d. UAU

8. Panagiotis is explaining to his nephew how he uses JavaScript at work. Which of the following would Panagiotis NOT say about JavaScript?
   a. JavaScript can create separate "stand-alone" applications.
   b. JavaScript instructions are embedded inside HTML documents.
   c. JavaScript interacts with the HTML page's Document Object Model (DOM).
   d. An HTML document that contains JavaScript code is downloaded onto the user's device.

9. Which of the following is true about a malvertising attack?
   a. Attackers directly infect the website that is being compromised by identifying a vulnerability in the web server.
   b. Java applets are attached to spam messages that pretend to be advertisements.
   c. Attackers may infect the third-party advertising networks so that their malware is distributed through ads sent to users' web browsers.
   d. Resource objects are sent as email attachments with a source that pretends to be a well-known advertising agency.

10. Christos's computer was infected by a drive-by download attack. What did Christos do to get infected?
   a. He clicked *Download*.
   b. He opened an infected email attachment.
   c. He viewed a website.
   d. He unknowingly sent a virus to a website.

11. Which of the following expands the normal capabilities of a web browser?
   a. Plug-on
   b. Extension
   c. Addition
   d. Module

12. What technique do attackers use to circumvent text-based spam filters?
   a. Object spam
   b. Attachment spam
   c. Flash spam
   d. Image spam

13. Which of the following is NOT an email risk?
   a. Encryption
   b. Embedded hyperlinks
   c. Spam
   d. Attachments

14. **What is the first step in defending against Internet-based attacks?**
- **a.** Add security extensions to the web browser.
- **b.** Use a web browser that supports automatic downloads.
- **c.** Never open any email attachments.
- **d.** Ensure that the device itself is properly secured.

15. **Why should you not click an embedded hyperlink in an email message?**
- **a.** Embedded hyperlinks are known to be flawed and can cause computer problems.
- **b.** Embedded hyperlinks take up too much disk space on your computer.
- **c.** Embedded hyperlinks can take you to a different website than what is being advertised.
- **d.** Embedded hyperlinks are slow.

16. **Which of these is true regarding malvertising?**
- **a.** Preventing malvertising can be accomplished through coordination between advertisers and their websites.
- **b.** Malvertising allows attackers to narrowly target their victims.
- **c.** Malvertising rarely occurs on big-name websites.
- **d.** Although popular at one time, malvertising is rarely used by attackers today because of increased scrutiny by security professionals.

17. **Which of the following attacks takes advantage of an authentication "token" that a website sends to a user's web browser?**
- **a.** CSRF
- **b.** XSS
- **c.** CRM
- **d.** AXAR

18. **Which of the following attacks has the goal of either eavesdropping on the conversation or impersonating one of the parties?**
- **a.** MIRX
- **b.** MITB
- **c.** Session replay
- **d.** MITM

19. **Why would you want to block external content from downloading into your email MAU client?**
- **a.** To take advantage of the remote reading pane.
- **b.** To slow down your email client so you can read the message.
- **c.** To prevent your computer's graphics processor buffer from filling too quickly.
- **d.** To prevent spammers from knowing that your email address is valid.

20. **Gabrielle wants to use a Microsoft Office feature that opens attachments in a read-only mode. Which feature would she choose?**
- **a.** Protected View
- **b.** Trusted documents
- **c.** Trusted location
- **d.** Trusted messages

Hands-On Projects

Project 4-1: Testing Web Browser Security

One of the first steps in securing a web browser is to analyze whether it has any security vulnerabilities. These vulnerabilities may be a result of missing patches or vulnerable extensions. In this project, you will test the security of a web browser.

1. Open the web browser Chrome on Microsoft Windows or Apple macOS.

Note 24

This security check can be performed using Firefox or Chrome on Microsoft Windows or Safari, Chrome, or Firefox on Apple macOS. The instructions for this project are for Chrome. There may be slight variations if another browser is used instead.

2. Enter the URL **https://browsercheck.qualys.com/**. (If you are no longer able to access the site through the web address, use a search engine to search for "Qualys Browser Check.")
3. Click **Learn more about Qualys BrowserCheck**.
4. Read the information on this page about what the Qualys browser check plug-in will do.
5. Return to the home page.

Continued

6. Click **Install Plugin**.
7. Check the **I have read and accepted the Service User Agreement** box.
8. Click **Continue**.
9. You will now be redirected to the Chrome Store to install the extension. Click **OK**.
10. If Chrome blocks the pop-up window from displaying, click the **Pop-ups blocked** icon and click **Allow pop-ups and directs from https://browsercheck.qualsy.com**. Click **Done**. Click **OK** to reload the page, click the check box again, and then **Continue**.
11. At the Chrome Web Store, click **Add to Chrome**.
12. Read the information about what the Qualys BrowserCheck can do. Is this helpful information? Would most users understand it?
13. Click **Add extension**.
14. Do not click **Turn on sync**.
15. Click the **Extensions** icon in Chrome to read the information. If necessary, click **Here** to download the BrowserCheck host application.
16. After the download is complete, click the file and follow the instructions to install the application.
17. Restart Chrome.
18. After the plug-in is installed, a Qualys button appears at the top of the browser. Click the **Qualys** button.
19. A message appears indicating that you must also install the Qualys BrowserCheck host application. Follow the instructions to download the host application.
20. Start the host application when the download is complete. Follow the default instructions to install the host application.
21. When the host application installation is complete, restart your browser.
22. Click the **Qualys BrowserCheck** extension to start the scan.
23. A brief analysis of your browser's security will appear. Is this information helpful?
24. Under **Scan Type**, change the selection to **Advanced Scan** to scan other software.
25. Click **Re-Scan**. Note that this may take several minutes to complete.
26. Click each tab in the **Results** section.
27. If any security issues are detected, click the **Fix It** button to correct each problem. Follow the instructions on each page to correct the problems.
28. Return to the Qualys scan results page.
29. Is this helpful software? Why or why not?
30. Close all windows.

Project 4-2: Chrome Web Browser Security Settings

In this project, you will configure several security settings for the Google Chrome web browser and assess the impact of the settings.

1. Launch the Google Chrome web browser.
2. Check to determine if Chrome is up to date. Click the "hamburger" icon and then click **Help** and **About Google Chrome**.
3. If the message **Update Google Chrome** appears, then Chrome needs to be updated. Click the button to update Chrome.
4. Close this tab.
5. Click the "hamburger" icon and then click **Settings**.
6. In the left pane, click **Privacy and security**.
7. Begin by performing an overall assessment. Under **Safety check**, click **Check now**. An analysis of the safety of Chrome will be displayed.
8. Click the arrow next to **Device software**.
9. Chrome can also scan for other malware on the computer. Click the **Find** button. The scan may take several minutes to run.
10. When completed, click the left arrow.

11. Next, examine the extensions. Click the icon next to **Extensions** to open a new tab that displays the installed extensions.
12. Review this list. If any extensions are no longer being used, click the **Remove** button.
13. Observe the permissions for an extension. Select an extension and click **Details**. Do the privileges that the permission has surprise you?
14. Click the **Back** arrow and then review the details of another extension. When finished, click the **Back** arrow.
15. Make sure that the most up-to-date version of these extensions has been installed. On the **Extensions** tab, click **Update** to update the extensions.
16. When the extension update is complete, close this tab to return to the Chrome **Settings** tab.
17. Review and set the security mode for the browser. Under **Privacy and Security**, click the right arrow next to **Security**.
18. Under **Safe Browsing**, note that three modes are available. If necessary, click **Enhanced protection**.
19. Under **Advanced**, make sure that **Always use secure connections** is turned on. This will create a warning for any site that is accessed that does not support HTTP.
20. Click the left arrow to return to the **Privacy and security** settings page.
21. Under **Privacy and security**, click the right arrow next to **Site Settings**.
22. Under **Content**, click the right arrow next to **JavaScript**.
23. Because JavaScript is considered a security risk, experiment by turning it off to determine if you could use Chrome without JavaScript. Click **Don't allow sites to use Javascript**.

Note 25

If it is too restrictive to not use JavaScript, then note the sites you frequent that require JavaScript and add them under **Allowed to use Javascript**. This will allow you to keep **Don't allow sites to use Javascript** turned on for other sites.

24. How difficult or easy is it to set Chrome security settings?
25. Close all windows.

Project 4-3: Edge Web Browser Security Settings

In this project, you will configure several security settings for the Microsoft Edge web browser.

1. Launch the Microsoft Edge web browser.
2. Check to determine if Edge is up to date. Click the horizontal ellipsis icon and then click **Help and feedback** and **About Microsoft Edge**.
3. If the **About** page shows **An update is available. Select Download and install to proceed. Network charges may apply** then select **Download and install**. It may be necessary to restart Edge.
4. In the left pane, click **Privacy, search, and services**.
5. Scroll down to the **Security** section.
6. Note that several options are available for enhanced security. At a minimum, be sure that the following settings are turned on:
   - **Microsoft Defender SmartScreen**: Helps protect from malicious sites and downloads
   - **Block potentially unwanted apps**: Blocks downloads of low-reputation apps that might cause unexpected behaviors
   - **Typosquatting Checker**: Warns if a mistyped URL may be directed to a potentially malicious site
   - **Turn on site safety services to get more info about the sites you visit**: Provides more site information when hovering over the padlock in the address bar
7. Scroll down to **Enhance your security on the web**. Note that there are two modes, *Balanced* and *Strict*. If necessary, click **Strict** to turn it on.
8. Click the right arrow next to **Exceptions**. This is the location where you can enter sites that are an exception to this level of security. Return to the **Settings** screen.

Continued

9. Click **Downloads** in the left pane.
10. Another level of enhanced security is **Open Office files in the browser**. This will allow you to view a document before downloading it to your device and potentially open the door for malware. Be sure this option is turned on.
11. How would you rate Edge security compared to Chrome security? Which is more comprehensive? Which has easier settings?
12. Close all windows.

Case Projects

Case Project 4-1: Comparing Web Browser Security Features

Of the most popular web browsers—Chrome, Edge, Firefox, and Safari—which has the best security features? Using the Internet, research the security features of each of these browsers. Create a table that lists the different security features. In your opinion, does one browser have more security features than the rest? Does a browser have fewer features? Give reasons for your conclusion and write a one-paragraph summary of your research.

Case Project 4-2: Alternative Web Browsers

Even though most users opt to use one of the four major web browsers (Chrome, Edge, Firefox, and Safari), other web browsers are available, and several of these are often promoted as having stronger security. Use the Internet to search for alternative secure web browsers. Identify three of these browsers and read information about their features. Download the browsers and test them. (You can do this in your Windows Sandbox.) What do you think about these web browsers? Do they provide enhanced security? Why or why not? Would you use them? Would you recommend them to others? Create a one-page summary of your work.

Case Project 4-3: Secure Email Protocols

Secure email protocols can help protect email messages. The *Secure/Multipurpose Internet Mail Extensions (S/MIME)* is a protocol for securing email messages. The *Secure Real-time Transport Protocol (SRTP)* is similar to S/MIME but also is designed to protect other types of communications. Use the Internet to research S/MIME and SRTP. What are their advantages and disadvantages? What protections do they provide? How easy or difficult are they to use? Write a one-page paper about what you have learned.

Case Project 4-4: Web Browser Extensions

Examine the web browser extensions that you have installed. Which do you use regularly? Which do you rarely use? Next, go to the "store" that supports your web browser and find one extension that would be helpful to you. Read the reviews about its features. Next, download and install the extension. How easy is it to install a web browser extension? Is it so easy that users could indiscriminately install too many extensions without understanding the risks? How could a web browser convey information to users about the risks of extensions? Should the browser require users to review their browser extensions every 90 days? Write a short paper on your research and how web browsers can help users understand the risks of extensions.

References

1. Boddy, Matt, "Exposed: Cyberattacks on Cloud Honeypots," *Sophos*, Apr. 9, 2019, accessed June 5, 2019, www.sophos.com/en-us/press-office/press-releases/2019/04/cybercriminals-attack-cloud-server-honeypot -within-52-seconds.aspx.
2. "Email & Spam Data," *Cisco Talos,* accessed July 7, 2020, https://talosintelligence.com/reputation_center /email_rep.
3. Thompson, Mia, "How to Stay Safe from Office Macro-based Malware with Email Security," *Solarwinds MSP*, Feb. 10, 2020, accessed June 17, 2020, https://www.solarwindsmsp.com/blog/how-stay-safe-office-macro -based-malware-email-security.
4. McCart, Craig, "15 Malvertising Statistics," *Comparitech*, Mar. 29, 2022, accessed May 10, 2022, https:// www.comparitech.com/blog/information-security/malvertising-statistics/#:~:text=1.,the%20COVID% 2019%20pandemic%20began.

Module 5

Mobile Security

After completing this module, you should be able to do the following:

1. Explain how Wi-Fi, Bluetooth, and Near Field Communication operate.
2. Identify attacks on wireless networks.
3. Describe different types of mobile devices.
4. Describe the risks associated with mobile devices.
5. Explain how to implement mobile defenses.

Cybersecurity Headlines

It's no surprise that apps on mobile devices that are used to handle our money are extremely popular today. These include apps used to check bank balances, buy stocks, and trade cryptocurrency. Mobile users install 4.6 billion finance apps globally each year. Users spend 16.3 billion hours in those apps, a 15 percent increase from year to year. Almost 9 out of every 10 Americans use a mobile device to check their bank balance. One bank reported that it has almost 57 million "digitally active customers," an increase of 10 percent from the previous year. Of these, almost 43 million customers were using its mobile app.

Recently, security researchers examined 400 of the most popular financial mobile apps available through both Apple and Google. The review included mobile apps used for banking, stock trading, portfolio management, insurance, accessing credit agencies, and cryptocurrency. They scored the mobile apps on a scale of 0–100 and also assigned a pass or fail letter grade based on the following criteria:

A (100–90) or B (89–80): Mobile apps that scored in this range represent high-quality, low-risk apps and are considered the most secure.

C (79–70): C-level apps have medium risk and should be used with caution and monitored for strange activity or scores changing with updates. Mobile apps may leak sensitive information or have excessive permissions that are unnecessary, such as a budgeting app that gains permission to access a contact address book, GPS data, or a camera.

D (69–60) or F (59 or less): These apps represent a high risk. They may leak unencrypted user IDs, passwords, or account information over the network or are open to man-in-the-middle attacks. They should not be used until security vulnerabilities are fixed by their developers.

The results of the research into the security and privacy of financial mobile apps were very eye-opening.

Most of the 400 finance-related apps failed to fully protect user security and privacy. Out of 400, a "remarkable" 263 (66 percent) scored a D or an F in security and privacy by having at least two high-risk vulnerabilities that leak sensitive data or leave users vulnerable to network attacks. Of these 263 failed apps, 15 percent contained a critical bug in an outdated third-party library, as well as at least one other critical flaw that allows attackers to collect or modify data through insecure Internet connections. And 11 of these failed apps scored 60 or below, containing at least two high-risk vulnerabilities that could be "devastating to users of a financially-regulated business and the business itself." Only 114 (29 percent) passed with a C grade, and just 23 (6 percent) of the 400 apps scored an A or B.

Continued

The researchers also looked at cryptocurrency apps. And the results were equally surprising.

The downloads of cryptocurrency-related mobile apps have grown dramatically, with one cryptocurrency wallet app reaching over 70 million users. The popular exchange Coinbase offers its mobile app to over 62 million token holders. There are many more cryptocurrency-related apps than there are mobile banking or stock trading apps, and these are the fastest-growing subset in the finance category.

Of the 250 popular cryptocurrency-related applications including wallets, exchanges, portfolio trackers, and cryptocurrency news apps, 71 percent (191) had a score of 59 or below. One cryptocurrency app scored only 6 out of 100 possible points. And only 16 cryptocurrency apps (6 percent) scored an A or B.

When it comes to using a secure mobile app to manage our finances, we still have an awfully long way to go.

The word *ubiquitous* means "ever-present" or "found everywhere." And that is an accurate description of wireless data networks and the mobile devices that connect to them. Thanks to our smartphones, tablets, laptop computers, and the wireless networks behind them, it is no longer necessary for us to use a desktop computer anchored by a cable to a network connection in a wall. Instead, we can use our mobile devices to freely communicate with others, surf the web for information, check email, or perform virtually any technology task from almost anywhere.

Statistics confirm the popularity of wireless networks and mobile devices. Over 97 percent of Americans are mobile phone users. Half of all web traffic is generated through a mobile device.[1] Seventy percent of users check product reviews on their smartphones before purchasing while almost two-thirds of smartphone usage is for watching videos. About 87 percent of users check their phones before going to sleep, 67 percent check their phones as soon as they wake up, and to save time, 12 percent of adults use a smartphone while showering. The popularity of mobile devices has led to a new dictionary word: *nomophobia* is the fear of not being with your mobile phone![2]

But just as users have flocked to mobile devices and wireless networks, so too have attackers. Mobile devices have seen an increase in malware and attacks directed at them. Wireless data networks have also become a prime target for attackers who attempt to capture the unprotected wireless signal freely floating through the air to uncover passwords, credit card numbers, and other important information. It is critical today to protect our mobile devices and wireless networks.

In this module, you will examine some of the attacks on mobile devices and the wireless data networks that support them. First, you will explore the types of attacks that a wireless network faces along with the attacks directed at mobile devices using these networks. Then, you will learn how to protect wireless networks and mobile devices.

Mobile Attacks

Several types of attacks are directed toward mobile devices. Understanding the attacks directed toward wireless networks used by these mobile devices is equally important.

Attacks on Wireless Networks

Three major types of wireless networks are popular today both among users as well as attackers. These networks are Wi-Fi, Bluetooth, and Near Field Communication.

Wi-Fi Networks

Wi-Fi networks have become commonplace today. Understanding what Wi-Fi is, the equipment needed to operate on a Wi-Fi network, and the attacks that this type of network faces are all important.

What Is Wi-Fi? **Wi-Fi** is a wireless data network technology that provides high-speed data connections for mobile devices. This type of network is technically known as a *wireless local area network* (*WLAN*) and is intended to replace or supplement a *wired local area network* (*LAN*). Devices such as tablets, laptop computers, smartphones, and wireless printers that are within range of a centrally located connection device can send and receive information at varying transmission speeds using radio frequency (RF) transmissions.

Caution 

Wi-Fi networks are different from cellular telephony networks that are designed, installed, and maintained by wireless telecommunication carriers. Cellular networks use standards such as *5G* and *4G LTE* for both voice and data communications and charge users accordingly for this coverage. Wi-Fi networks, in comparison, are set up and maintained by users. They are generally faster than cellular telephony networks although they have a smaller geographical area of coverage. Almost all mobile devices can switch between using cellular telephony networks and Wi-Fi.

In the field of computer networking and wireless communications, the most widely known and influential organization is the **Institute of Electrical and Electronics Engineers (IEEE)**. The IEEE and its predecessor organizations date back to 1884. The IEEE is one of the leading developers of global standards in a broad range of industries such as energy, transportation, biomedical, and healthcare, and is currently involved in developing and revising over 800 standards.

Note 1

Although some publications and websites may claim that Wi-Fi stands for *wireless fidelity*, it is not true. Instead, *Wi-Fi* was created by a marketing firm because the wireless industry was looking for a user-friendly name instead of *IEEE* or *WLAN*. Wi-Fi is considered a proper noun and is a registered trademark, and variations (*WiFi*, *wifi*, or *Wifi*) should not be used as alternatives.

For over 30 years, the IEEE has been responsible for establishing standards for Wi-Fi networks. As wireless technology has matured, new standards have been introduced that primarily address faster speeds. As these new standards are ratified, wireless hardware vendors respond by producing and selling ever-faster Wi-Fi devices. The IEEE uses a sometimes-confusing lettering system for designating these successive standards, such as *IEEE 802.11n* and *IEEE 802.11ac*. To reduce confusion, in 2018 the Wi-Fi Alliance adopted "consumer-friendly" version numbers instead of using the IEEE nomenclature. Table 5-1 compares the different Wi-Fi names and standards. The speed (*data rates*) is measured in *megabits* or millions of bits per second (Mbps) and *gigabits* or billions of bits per second (Gbps).

Table 5-1 Wi-Fi names and standards

Wi-Fi Alliance version	IEEE name	Ratification date	Frequency utilized	Maximum data rate
None	802.11	1997	2.4 GHz	2 Mbps
Wi-Fi 1	802.11b	1999	2.4 GHz	11 Mbps
Wi-Fi 2	802.11a	1999	5 GHz	54 Mbps
Wi-Fi 3	802.11g	2003	2.4 GHz	54 Mbps
Wi-Fi 4	802.11n	2009	2.4 GHz & 5 GHz	600 Mbps
Wi-Fi 5	802.11ac	2014	5 GHz	7.2 Gbps
Wi-Fi 6	802.11ax	2019	2.4 GHz & 5 GHz	9.6 Gbps
Wi-Fi 6E	802.11ax	2020	1-6 GHz	9.6 Gbps

Note 2

The next version of Wi-Fi will be called Wi-Fi 7. It will offer a maximum data rate of 33 Gbps and may be ratified by 2023.

There is a trade-off between *speed* and *distance* with wireless networks: the *faster* the speed, the *closer* the devices must be to the centrally located connection device. That means as Wi-Fi networks have become faster, they have not correspondingly increased the distance between devices. For example, the speed of a Wi-Fi 4 network is much slower than a Wi-Fi 5 network (600 Mbps vs. 7.2 Gbps) but Wi-Fi 4 devices have a range of 230 feet (70 meters) while Wi-Fi 5 devices have only a range of 115 feet (35 meters).

Wi-Fi Equipment For all of its functionality, the list of necessary equipment for a Wi-Fi network to operate is surprisingly short. First, each mobile device must have a means to send and receive the wireless signals. Originally laptops and desktop computers required a separate apparatus with an antenna called a **wireless adapter**. Today these are internal to mobile devices (called a Wi-Fi "chipset") with the antenna either embedded into the chipset itself or as a small wire around the frame of the mobile device.

Note 3

The antenna wire for a laptop is embedded above the screen because this location provides the best reception and minimizes potential interference of the RF signal being blocked by obstructions.

Second, a Wi-Fi network needs a centrally located controller Wi-Fi device. This device primarily consists of an antenna and a radio transmitter/receiver to send and receive wireless signals, special bridging software to interface wireless devices to other devices, and a wired network interface that allows it to connect by cable to a standard wired network.

A centrally located Wi-Fi device for a home is called a *residential WLAN gateway* as it is the entry point ("gateway") from the Internet into the Wi-Fi network. However, most vendors instead choose to label their products as *wireless broadband routers* or more simply **wireless routers**. The wireless router acts as the "base station" for the wireless devices, sending and receiving wireless signals between all devices as well as providing access to the external Internet. The wireless router is either connected to the modem that is in turn attached to an Internet connection or the wireless router is part of the modem itself. A wireless router is illustrated in Figure 5-1.

Figure 5-1 Wireless router

Sergey Gostev/Shutterstock.com

Note 4

A wireless router combines several networking technologies and even includes an external firewall for additional protection.

A home Wi-Fi network is shown in Figure 5-2. The wireless transmission signal from the wireless router allows the laptop, smartphone, and printer to wirelessly connect to it, forming the Wi-Fi network. The wireless router in turn is connected to the modem, which also connects to both the wired network as well as the Internet. This enables all the devices to equally share resources: the wireless laptop can use the wired scanner, just as the wired personal computer can access the wireless printer. And all wired and wireless devices can share a single Internet connection.

Figure 5-2 Home Wi-Fi network

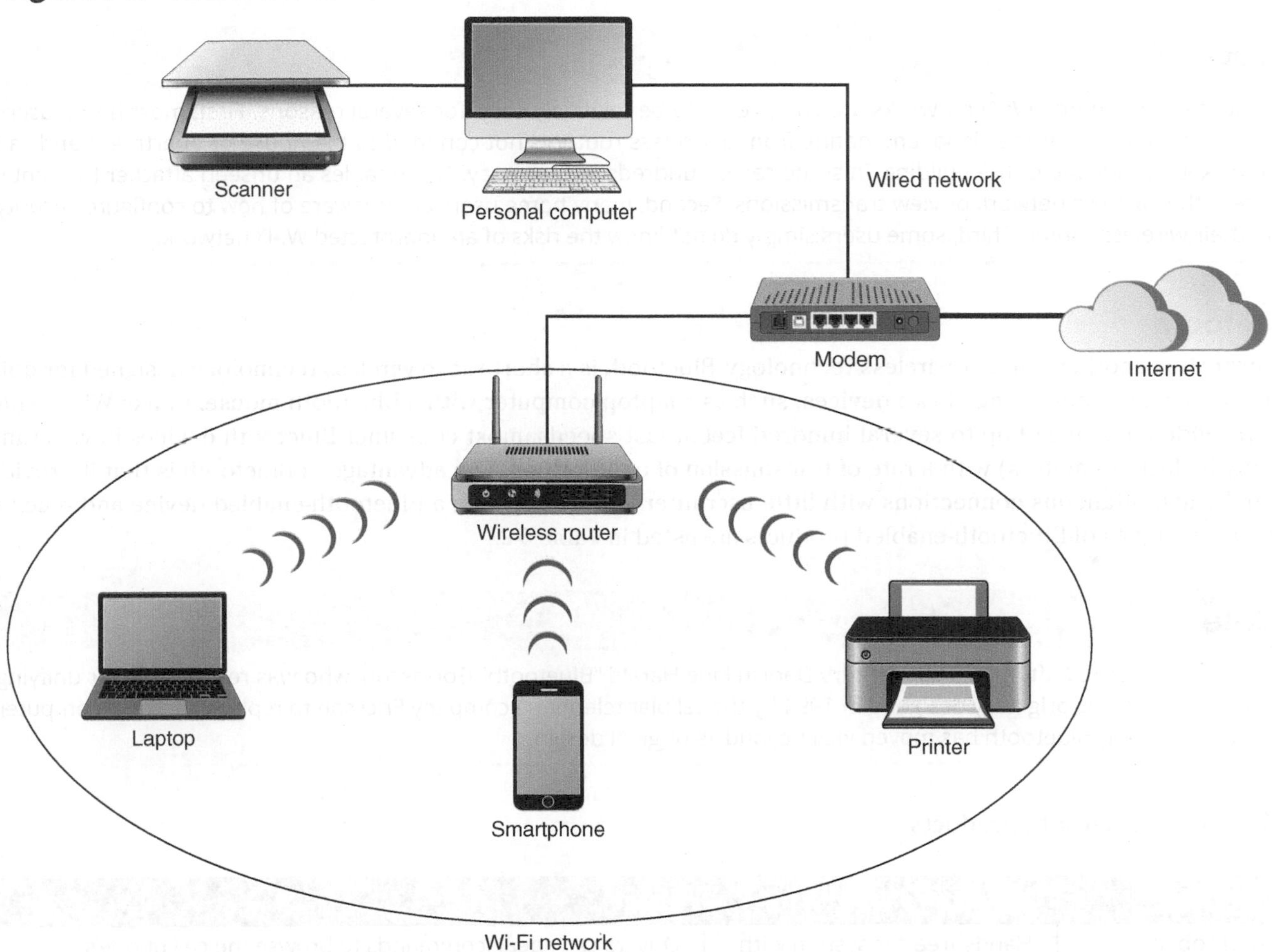

A new trend in home Wi-Fi helps address slower data transmissions. Obstacles such as walls can slow a Wi-Fi signal, and a device that is farther away from the wireless router also has a slower connection. A **mesh** Wi-Fi network consists of a main wireless router that connects directly to the modem along with a series of "satellite" modules called *nodes* that are placed around the house. Wireless devices automatically connect to the closest node. This helps reduce the impact of obstacles and extend the geographical area in which a faster signal can be received.

In a business or school setting, instead of using a single wireless router, a more sophisticated device known as an **access point (AP)** is used. Businesses typically have multiple APs, often hundreds or even thousands spread throughout a building or across a campus. Because the wireless signal can only be transmitted for several hundred feet, multiple APs are used to provide "cells" or areas of coverage. As the user moves (called *roaming*) from one cell to another with their wireless device, a *handoff* occurs so that the AP to which the user is closest now becomes the new base station.

Attacks on Wi-Fi Users face several risks from attacks on home Wi-Fi networks. Among other things, attackers can do the following:

- **Steal data**. On a computer in the home Wi-Fi network, an attacker could access any folder with file sharing enabled. This essentially provides an attacker full access to steal sensitive data from the computer.
- **Read wireless transmissions**. User names, passwords, credit card numbers, and other information sent over the wireless network could be captured by an attacker.
- **Inject malware**. Because attackers could access the network behind a firewall, they could inject viruses and other malware onto the computer.
- **Download harmful content**. In several instances, attackers have accessed a home computer through an unprotected Wi-Fi network and downloaded child pornography to the computer, and then turned that computer into a file server to distribute the content. When authorities have traced the files back to that computer, the unsuspecting owner has been arrested and his equipment confiscated.

Note 5

Attacks against home Wi-Fi networks are considered to be relatively easy, for several reasons. First, most home users overlook the fact that the signal emanating from a wireless router is not confined to the house or apartment and can be picked up outside of the building, in some cases hundreds of feet away. This enables an unseen attacker to silently access the wireless network or view transmissions. Second, many home users are unaware of how to configure security on their wireless router. Third, some users simply do not know the risks of an unprotected Wi-Fi network.

Bluetooth

Bluetooth is another popular wireless technology. Bluetooth is a short-range wireless technology designed for quick "pairing" or interconnecting of two devices, such as a laptop computer with a Bluetooth mouse. Unlike Wi-Fi, which can provide coverage of up to several hundred feet at fast speeds, most consumer Bluetooth devices have a range of only 33 feet (10 meters) with a rate of transmission of only 1 Mbps. The advantage of Bluetooth is that it provides virtually instantaneous connections with little user intervention between a Bluetooth-enabled device and receiver. Several examples of Bluetooth-enabled products are listed in Table 5-2.

Note 6

Bluetooth is named after the 10th-century Danish King Harald "Bluetooth" Gormsson, who was responsible for unifying Scandinavia. It was originally designed in 1994 by the cellular telephone company Ericsson to replace personal computer cables. However, Bluetooth has moved well beyond its original design.

Table 5-2 Bluetooth products

Category	Bluetooth pairing	Usage
Automobile	Hands-free car system with cell phone	Drivers can speak commands to browse the cell phone's contact list, make and receive hands-free phone calls, or use its navigation system.
Home entertainment	Stereo headphones with portable music player	Users can create a playlist on a portable music player and listen through a set of wireless headphones or speakers.
Photographs	Digital camera with printer	Digital photos can be sent directly to a photo printer or from pictures taken on one cell phone to another phone.
Computer accessories	Computer with keyboard and mouse	A small travel mouse can be linked to a laptop or a full-size mouse and keyboard can be connected to a desktop computer.
Sports and fitness	Heart rate monitor with wristwatch	Exercisers can track heart rates and blood oxygen levels.
Medical and health	Blood pressure monitors with smartphones	Patient information can be sent to a smartphone, which can then send an emergency phone message if necessary.

Note 7

Bluetooth is also finding its way into some unlikely devices. A Victorinox Swiss Army pocketknife model has Bluetooth technology that can be used to remotely control a computer when projecting a PowerPoint presentation. Other unusual devices include Bluetooth-enabled toothbrushes, keychain breathalyzers, stethoscopes, and even trash cans that send reminders to take out the garbage.

Because of the "on-the-fly" nature of Bluetooth pairings, attacks on wireless Bluetooth technology are not uncommon. Two Bluetooth attacks are bluejacking and bluesnarfing. **Bluejacking** is an attack that sends unsolicited messages to Bluetooth-enabled devices. Usually bluejacking involves sending text messages, but images and sounds can also be transmitted. Bluejacking is usually considered more annoying than harmful because no data is stolen. However, many Bluetooth users resent receiving unsolicited messages. **Bluesnarfing** is an attack that accesses unauthorized information from a wireless device through a Bluetooth connection, often between cell phones and laptop computers. In a bluesnarfing attack, the attacker copies emails, calendars, contact lists, cell phone pictures, or videos by connecting to the Bluetooth device without the owner's knowledge or permission.

Near Field Communication

Near field communication (NFC) is a set of standards used to establish communication between devices in very close proximity. When the devices are brought within 4 centimeters of each other or tapped together, two-way communication is established. Devices using NFC can be active or passive. A *passive NFC* device, such as an NFC tag, contains information that other devices can read though the tag does not read or receive any information. *Active NFC* devices can read information as well as transmit data.

Examples of NFC use include the following:

- **Entertainment**. NFC devices can be used as a ticket to a stadium or concert, for purchasing food and beverages, and for downloading upcoming events by tapping a smart poster.
- **Office**. An NFC-enabled device can be used to enter an office, clock in and out on a factory floor, or purchase snacks from a vending machine.
- **Retail stores**. Coupons or customer reward cards can be provided by tapping the point-of-sale (PoS) terminal.
- **Transportation**. On a bus or train, NFC can be used to quickly pass through turnstiles and receive updated schedules by tapping the device on a kiosk.

Note 8

There are five types of NFC tags, Type 1 through Type 5, which are used in different settings. For example, Type 2 tags are often used for event tickets and transit passes while Type 5 is used to tag library books.

Consumer NFC devices are often used as an alternative to payment methods using cash or a credit card. These are called **contactless payment systems**. Users can either store payment card numbers in a "virtual wallet" on a watch or smartphone or use a contactless credit card with an NFC chip. This enables quick and touchless payments for purchases, as illustrated in Figure 5-3.

Figure 5-3 Contactless payment system

Note 9

About two out of every three retailers now offer contactless payment systems.

The use of NFC has risks because of the nature of this technology. The risks and defenses of using NFC are listed in Table 5-3.

Table 5-3 NFC risks and defenses

Vulnerability	Explanation	Defense
Eavesdropping	Unencrypted NFC communication between the device and terminal can be intercepted and viewed.	Because an attacker must be extremely close to pick up the signal, users should remain aware of their surroundings while making a payment.
Data theft	Attackers can "bump" a portable reader to a user's smartphone in a crowd to make an NFC connection and steal payment information stored on the phone.	This can be prevented by turning off NFC while in a large crowd.
Man-in-the-middle attack	An attacker can intercept the NFC communications between devices and forge a fictitious response.	Devices can be configured in pairing so one device can only send while the other can only receive.
Device theft	The theft of a smartphone could allow an attacker to use that phone for purchases.	Smartphones should be protected with passwords or strong PINs.

Attacks on Mobile Devices

At one time, the idea of having a mobile device that can be carried and used virtually anywhere seemed like "science fiction." However, today they are commonplace. There are different types of mobile devices, and each type of mobile device faces several cybersecurity risks.

Types of Mobile Devices

Most mobile devices have a common set of core features, which serve to differentiate them from other computing devices. Many (but not all) mobile devices also have additional features. These features are listed in Table 5-4.

Table 5-4 Mobile device core and additional features

Core features	Additional features
Small form factor	Global Positioning System (GPS)
Mobile operating system	Microphone and/or digital camera
Wireless data network interface for accessing the Internet, such as Wi-Fi or cellular telephony	Wireless cellular connection for voice communications
Applications (apps) that can be acquired through different means	Wireless personal area network interfaces like Bluetooth or near field communication (NFC)
Local non-removable data storage	Removable storage media
Data synchronization capabilities with a separate computer or remote servers	Support for using the device itself as removable storage for another computing device

Several types of mobile devices include tablets, smartphones, wearables, and portable computers.

Tablets **Tablets** are portable computing devices first introduced in 2010. Designed for user convenience, tablets are thinner, lighter, easier to carry, and more intuitive to use than other types of computers. Tablets are often classified by their screen size. The two most common categories of tablet screen sizes are 5–8.5 inches (12.7–21.5 cm) and 8.5–10 inches (21.5–25.4 cm). The weight of tablets is generally less than 1.5 pounds (0.68 kg), and they are less than 1/2 inch (1.2 cm) thick. Figure 5-4 shows a typical tablet device.

Figure 5-4 Tablet device

Tero Vesalainen/Shutterstock.com

Note 10

Tablets have a sensor called an accelerometer that senses vibrations and movements. It can determine the orientation of the device so that the screen image is always displayed upright.

Tablets lack a built-in keyboard or mouse. Instead, for user input, they rely on a touchscreen that is manipulated with touch gestures, such as "double tap," "flick," and "pinch open." Some tablets allow for a keyboard and mouse to be attached.

Although tablets are primarily display devices with limited computing power, they have proven to be popular. Besides their portability, a primary reason for their popularity is that tablet computers have an operating system (OS) that allows them to run third-party apps. The most popular OSs for tablets are Apple iOS, Google Android, and Microsoft Windows.

Smartphones Early cellular telephones were called *feature phones* because they included a limited number of features, such as a camera, an MP3 music player, and a tool for sending and receiving text messages. Many feature phones were designed to highlight a single feature, such as the ability to take high-quality photos or provide a large amount of memory for music storage.

The feature phone has given way to today's **smartphone**, which has all the tools of a feature phone but also includes an OS that allows it to run apps and access the Internet. Because it has an OS, a smartphone offers a broader range of functionality. Users can install apps that perform functions for productivity, social networking, music, and so forth, much like a standard computer.

Note 11

Because the OS gives them the ability to run apps, smartphones are essentially handheld personal computers.

Wearables Another class of mobile technology consists of devices that can be worn by the user instead of carried. Known as **wearables**, these devices can provide even greater flexibility and mobility.

The most popular wearable technology is a *smartwatch*. Early smartwatches were just a means to receive smartphone notifications on the user's wrist. However, contemporary wearables have significantly evolved into a much higher-level device. A modern smartwatch can receive notifications of a phone call or text message, be used as a contactless payment system, and even call emergency services if the watch detects the user has fallen. Figure 5-5 displays a smartwatch.

Figure 5-5 Smartwatch

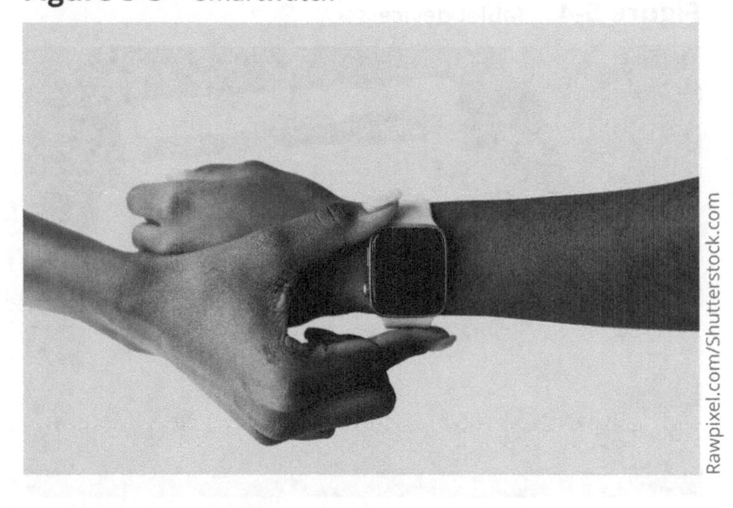

Rawpixel.com/Shutterstock.com

Another popular type of wearable is a *fitness tracker*. Originally designed to monitor and record physical activity, such as counting steps, fitness trackers likewise have evolved into sophisticated health monitoring devices. Modern fitness trackers can provide continuous heart rate monitoring, GPS tracking, oxygen consumption, repetition counting (for weight training), and sleep monitoring.

Portable Computers As a class, **portable computers** are devices that closely resemble standard desktop computers. Portable computers have similar hardware (keyboard, hard disk drive, RAM, etc.) and run the same OS (Windows, Apple macOS, or Linux) and application software (Microsoft Office, web browsers, etc.) that are found on a general-purpose computer. The primary difference is that portable computers are smaller self-contained devices that can easily be transported from one location to another while operating on battery power.

A *laptop* computer is regarded as the earliest portable computer. A laptop is designed to replicate the abilities of a desktop computer with only slightly less processing power yet is small enough to be used on a lap or small table. A *notebook* computer is a smaller version of a laptop computer and is considered a lightweight personal computer. Notebook computers typically weigh less than laptops and are small enough to fit easily inside a briefcase. A *subnotebook* computer is even smaller than standard notebooks and uses a low-power processor. A *2-in-1* computer (also called a *hybrid* or *convertible*) can be used as either a subnotebook or a tablet. They can be transformed from a subnotebook to a tablet through either a folding design or as a "slate" with a detachable keyboard, as seen in Figure 5-6.

Figure 5-6 2-in-1 computer with slate design

Andrey_Popov/Shutterstock.com

Note 12

One of the first mobile devices was a *personal digital assistant (PDA)*, a handheld mobile device intended to replace paper systems. Most PDAs had a touchscreen for entering data while others had a rudimentary keyboard that contained only a numeric keypad or thumb keyboard. Popular in the 1990s and early 2000s, PDAs fell out of favor as smartphones gained in popularity.

A popular type of computing device that resembles a laptop computer is a *web-based computer*. It contains a limited version of an OS and a web browser with an integrated media player. Web-based computers are designed to be used while connected to the Internet. No traditional software applications can be installed, and no user files are stored locally on the device. Instead, the device accesses online web apps and saves user files on the Internet. The most common OSs for web-based computers are the Google Chrome OS and Microsoft Windows 11 in S Mode.

Mobile Device Risks

Several vulnerabilities are inherent to mobile devices. The mobile device risks include installing unsecured applications, accessing untrusted content, limited physical security, constrained updates, connecting to public Wi-Fi, and location tracking.

Installing Unsecured Applications Normally users cannot download and install unapproved apps on their Apple iOS or Google Android mobile device. Apps can only be installed through the Apple App Store or Google Play Store (or another Android store); in fact, Apple devices can *only* download from the App Store.

However, users can circumvent the installed built-in limitations on their smartphone (called **jailbreaking** on Apple iOS devices or **rooting** on Android devices) to download from an unofficial third-party app store (called **sideloading**). Because these apps have not been vetted, they may contain security vulnerabilities or even malicious code. In addition, jailbreaking and rooting give access to the underlying OS and file system of the mobile device with full permissions, thus bypassing built-in security protections.

Note 13

In early 2022, the European Union (EU) proposed a regulation known as the Digital Markets Act (DMA) that would force Apple to make major changes to user installation of apps. Under DMA, Apple would be forced to allow users to install apps from other third-party sources (sideloading). It would also allow developers to use the App Store without using Apple's payment systems, which originally required developers to pay a fee of a 30 percent commission (this was reduced by Apple to 15 percent in early 2022). Apple argues that the DMA would weaken the security of their products, while the EU says it would allow smartphone owners to "also opt for other safe app stores."

Accessing Untrusted Content Another means by which untrusted content can invade mobile devices is through **short message service (SMS)**, which are text messages of a maximum of 160 characters, **multimedia messaging service (MMS)**, which provides for pictures, video, or audio to be included in text messages, or **rich communication services (RCS)**, which can convert a texting app into a live chat platform and supports pictures, videos, location, stickers, and emojis. Threat actors can send SMS messages that contain links to untrusted content or send a specially crafted MMS or RCS video that can introduce malware into the device.

Mobile devices also can access untrusted content that other types of computing devices generally do not face. One example is **Quick Response (QR)** codes. These codes are a matrix of two-dimensional barcodes that consist of black modules (square dots) arranged in a square grid on a white background. QR codes can store website URLs, plain text, phone numbers, email addresses, or virtually any alphanumeric data up to 4296 characters, which can be read by an imaging device such as a mobile device's camera. A QR code for www.cengage.com is illustrated in Figure 5-7.

Figure 5-7 QR code

Source: Qrstuff.com

An attacker can create an advertisement listing a reputable website, such as a bank, but include a QR code that contains a malicious URL. Once the user snaps a picture of the QR code using his mobile device's camera, the code directs the web browser on his mobile device to the attacker's imposter website or to a site that immediately downloads malware.

Limited Physical Security The greatest asset of a mobile device—its portability—is also one of its greatest vulnerabilities. Unlike desktop computers, mobile devices by their very nature are designed to be used in a wide variety of locations, both public (coffee shops, hotels, conference centers) and private (employee homes, cars). Yet these locations open the door for mobile devices to be misplaced or stolen. More than 8.7 million mobile phones are lost or stolen annually, which is over 24,000 phones each day.[3] Almost half of all laptop thefts occur from school offices and classrooms.[4]

Caution !

The risk of the loss of a mobile device is heightened if that device is used to access corporate business data. Unless properly protected, any data on a stolen or lost device could be retrieved by a thief. Also, the device itself can serve as an entry point into corporate data. On average, every employee at an organization has access to 17 million files and 1.21 million folders. The average organization has more than half a million sensitive files, and 17 percent of all sensitive files are accessible to each employee.[5]

Constrained Updates Currently, there are two dominant OSs for mobile devices. Apple iOS, developed by Apple for its mobile devices, is a closed and proprietary architecture. Google Android is not proprietary but is open for any original equipment manufacturer (OEM) to install or even modify. (However, modifications must adhere to Google's criteria in order to access all Google services.) Many OEMs worldwide make mobile devices that use Android because it is freely available.

Security patches and updates for these two mobile OSs are distributed through over-the-air (OTA) updates. Apple commits to providing OTA updates for up to eight years after the OS is released. However, OTA updates for Android OSs vary considerably. Mobile hardware devices developed and sold by Google receive Android OTA updates for three years after the device is first released. Other OEMs are required to provide OTAs for at least two years. However, after two years, many OEMs are hesitant to distribute Google updates because it limits their ability to differentiate themselves from competitors if all versions of Android start to look the same through updates. Also, because OEMs want to sell as many devices as possible, they have no financial incentive to update mobile devices that users would then continue to use indefinitely.

Whereas users once regularly purchased new mobile devices about every two years, that is no longer the case. Due to the high cost of some mobile devices, more users are keeping their devices for longer periods. This can result in a mobile device being used that is no longer receiving OTA security updates and thus has become vulnerable.

Connecting to Public Wi-Fi While using a mobile device outside its normal home Wi-Fi (in which the owner can set the security), users rely on public Wi-Fi for Internet access. Because these networks are beyond the control of the user, the type of security that is available may be suspect. If proper security is not implemented, attackers can eavesdrop on the data transmissions and view sensitive information.

In addition, an attacker may set up an **evil twin**. An evil twin is an AP or another computer designed to mimic an authorized Wi-Fi device. A user's mobile device may unknowingly connect to this evil twin instead of the authorized device so that attackers can receive any sensitive transmissions or directly send malware to the user's computer.

Location Tracking Virtually all modern smartphones support **geolocation**, or the process of identifying the geographical location of the device. By identifying the location of a person carrying a mobile device, the location of a close friend can be detected or the address of the nearest coffee shop can be displayed. Location services are used extensively by social media, navigation systems, weather systems, and other mobile-aware applications.

Note 14

Geolocation is based on the Global Positioning System (GPS), which is a satellite-based navigation system that provides information to a GPS receiver anywhere on (or near) the Earth where there is an unobstructed line of sight to four or more GPS satellites.

However, mobile devices using geolocation are at increased risk of targeted physical attacks. An attacker can determine where the user with the mobile device is currently located and use that information to follow the user to steal the mobile device or inflict harm upon the person. In addition, attackers can craft attacks by compiling a list of people with whom the user associates and the types of activities they perform.

A related risk is **GPS tagging** (also called *geotagging*), which is adding geographical identification data to media such as digital photos taken on a mobile device. A user who, for example, posts a photo on a social networking site may inadvertently be identifying a specific private location to anyone who can access the photo.

Two Rights & A Wrong

1. A wireless router serves as a base station for wireless devices, sending and receiving wireless signals between all devices as well as providing the access to the external Internet.
2. Bluetooth is a short-range wireless technology designed for the interconnection of two devices.
3. Downloading apps from an unofficial third-party app store is called jailbreaking.

See Appendix A for the answer.

Mobile Defenses

Although many attacks are directed at mobile devices and wireless networks, several defenses can protect against cyberattacks. These defenses can be classified as defenses for wireless networks and defenses for protecting wireless devices.

Wireless Network Security

Reducing the risk of attack through wireless networks is a necessary security practice. This involves securing a Wi-Fi home wireless network, following secure practices for using a public wireless network safely, and configuring Bluetooth on devices.

Home Wi-Fi Security

Configuring a Wi-Fi wireless router to provide the highest level of security protection is an important step. Configuring the router includes securing it and turning on Wi-Fi Protected Access Personal. In addition, other security settings can be implemented.

Securing the Wireless Router Securing a Wi-Fi wireless router is essential but is frequently overlooked. It involves "locking down" the device to prevent attackers from accessing it and changing the security settings. Locking down a wireless router means setting a strong default password, applying security patches, and disabling remote administration.

Set Strong Default Password The first step in securing a wireless router is to create a strong password to protect the router's internal configuration settings. Most wireless routers come preconfigured with a default password, and these passwords are advertised online and well known by attackers. Protecting the router with a strong password prevents attackers from remotely accessing the wireless router and turning off any security settings.

Note 15

A growing trend is to omit a default password on a wireless router. Instead, the OEM creates a unique password for each device. If the password is strong, it does not necessarily need to be changed; however, if the password is weak, it should be changed.

To secure the wireless router, the address of the router and the default password must first be known. This information is included in the documentation of the wireless router, can be obtained through the vendor's website, or can be determined by examining the network settings. The wireless router's Internet Protocol (IP) address, such as 192.168.1.254, can be entered into a web browser on a computer connected to the wireless network, which then displays the router's login screen. After the default password is entered, access is granted to the configuration settings of the router where the password can be changed.

Apply Security Patches Unlike computers and smartphones that are regularly patched with security upgrades, wireless routers may not regularly receive important patches from their OEMs. If a wireless router does not receive regular updates, it may be necessary to search the wireless router configuration settings to determine if they include an option to update the router. If that option exists, it should be checked periodically to determine whether an update is available. As an alternative, the OEM website that supports the router may contain information about a new update that can be downloaded and applied.

Note 16

A recent study of 127 new home routers from a variety of brands and models showed just how weak consumer router security is. Of the 127 routers examined, *none* of the routers was secure. About one out of every three of these routers used a version of the software that was released over 10 years ago and has 233 known security vulnerabilities. Only 81 of the wireless routers received one security patch update in the last year, 22 devices were not updated within the last two years, and one device was not updated within the last five years. In another study, a researcher analyzed every consumer wireless router for sale at a typical store. None of these had received a patch update in the last 14 months.

Disable Remote Administration Many routers also permit remote management of the router's configuration settings through the Internet. There are several router configuration options for remote management. The typical settings, as illustrated in Figure 5-8, are as follows:

Figure 5-8 Wireless router remote access settings

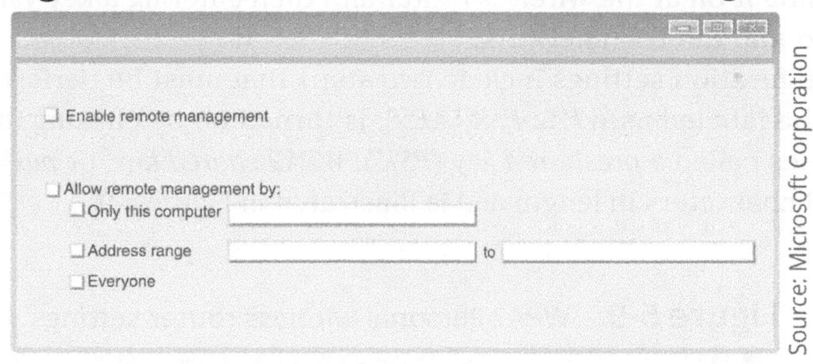

Source: Microsoft Corporation

- **Enable remote management**. This setting permits users to access the router's configuration settings from another location through the Internet.
- **Allow remote management by**. This option designates which devices can perform remote management. It can be one device (*Only this computer*), multiple devices (*Address range*), or all devices (*Everyone*).

Note 17

Some newer wireless routers provide remote management and wireless network access via an app installed on a smartphone or tablet. This app lets users check to see if a computer, mobile device, gaming console, media player, or other device is attached to the wireless network. In addition, email alert notifications can be sent to warn owners of a security intrusion attempt into the network or whenever a security update is available.

Unless remote management is essential, it is recommended that this feature be disabled. Turning remote management off adds a stronger degree of security because it limits access to the configuration settings of the wireless router to only the local computer connected to it.

Note 18

Despite the critical importance of protecting a home Wi-Fi wireless router, most users fail to do so. Of over 2,200 users surveyed, 82 percent said they had never changed their router's default administrative password, 86 percent had never updated the router, and 70 percent had never checked to see if any unknown devices were on their networks. More than half of those surveyed (51 percent) had never done any of these, and 48 percent said they did not understand why they would even need to.[6]

Turning On Wi-Fi Protected Access Personal The wireless signal that comes from a wireless router can be picked up outside of the building where the router is located, in some cases hundreds of feet away. On an unsecured wireless network, virtually anyone can access the Wi-Fi signal to read transmissions and access the network. This means two key security tasks must be performed: to mask the transmission so that no one can read any information being sent and received, and to prevent unauthorized users from accessing the network.

Although these two tasks may seem daunting, they can easily be accomplished by turning on *Wi-Fi Protected Access Personal*. This provides the optimum level of wireless security and is part of all certified wireless devices.

There are two versions of Wi-Fi Protected Access Personal, Wi-Fi Protected Access 2 (WPA2) Personal and Wi-Fi Protected Access 3 (WPA3) Personal.

Wi-Fi Protected Access 2 (WPA2) Personal For all devices other than Wi-Fi 6E devices, **Wi-Fi Protected Access 2 (WPA2) Personal** provides a high level of wireless security and is part of all certified wireless devices. Implementing WPA2 Personal involves turning it on at the wireless router and then entering a key value on each authorized device that has been preapproved to join the Wi-Fi network.

The wireless router configuration settings include two steps that must be performed. First, the WPA2 Personal security option, which may be labeled as *WPA2-PSK* [*AES*], is turned on by clicking the appropriate option button. Second, a key value, sometimes called a *preshared key* (*PSK*), *WPA2 shared key*, or *passphrase*, must be entered. This key value can be from 8 to 63 characters in length and is illustrated in Figure 5-9.

Figure 5-9 WPA2 Personal wireless router settings

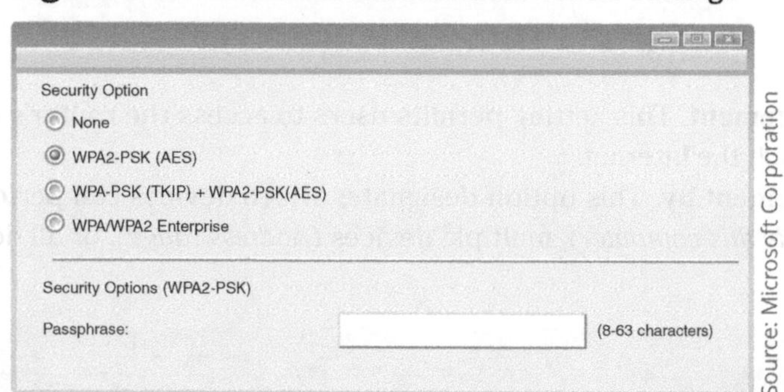

After enabling WPA2 Personal on the wireless router and entering a key value, the same key value must also be entered on each mobile device that has permission to access the Wi-Fi network. A mobile device that attempts to access a wireless network with WPA2 Personal will automatically ask for the key value. After this value is entered, the device can access the wireless network. Once the key value is entered, the mobile device can retain the value and does need to ask for it again.

To simplify WPA2 Personal, many wireless routers support **Wi-Fi Protected Setup (WPS)** as an optional means of configuring security. There are two common WPS methods. The PIN method uses a personal identification number (PIN) printed on a label on the wireless router or displayed through a software setup wizard. The user enters the PIN on the mobile device (such as a tablet or laptop computer), and the security configuration automatically occurs. The second method is the push-button method: the user pushes a button (usually a physical button on the wireless router and a virtual one displayed through a software setup wizard on the wireless device), and the security configuration takes place. However, it has been revealed that the PIN method in WPS has significant security design and implementation flaws. It is recommended that only the manual configuration method be used or that WPS be disabled on the wireless router and then turned on only temporarily when adding a new device to the Wi-Fi network.

Wi-Fi Protected Access 3 (WPA3) Personal For devices based on Wi-Fi 6E, **Wi-Fi Protected Access 3 (WPA3) Personal** provides the optimum level of wireless security. Like implementing WPA3 Personal, this involves turning it on at the wireless router, as seen in Figure 5-10. However, unlike WPA2, it is not necessary to enter a key value on each authorized device that has been preapproved to join the Wi-Fi network. Rather than relying on shared passwords, WPA3 "signs up" devices through Wi-Fi Device Provisioning Protocol (DPP). Users use QR codes or NFC tags to approve devices to join the network. By snapping a picture or receiving a radio signal from the wireless router, a device can be authenticated to the Wi-Fi network.

Figure 5-10 WPA3 Personal wireless router settings

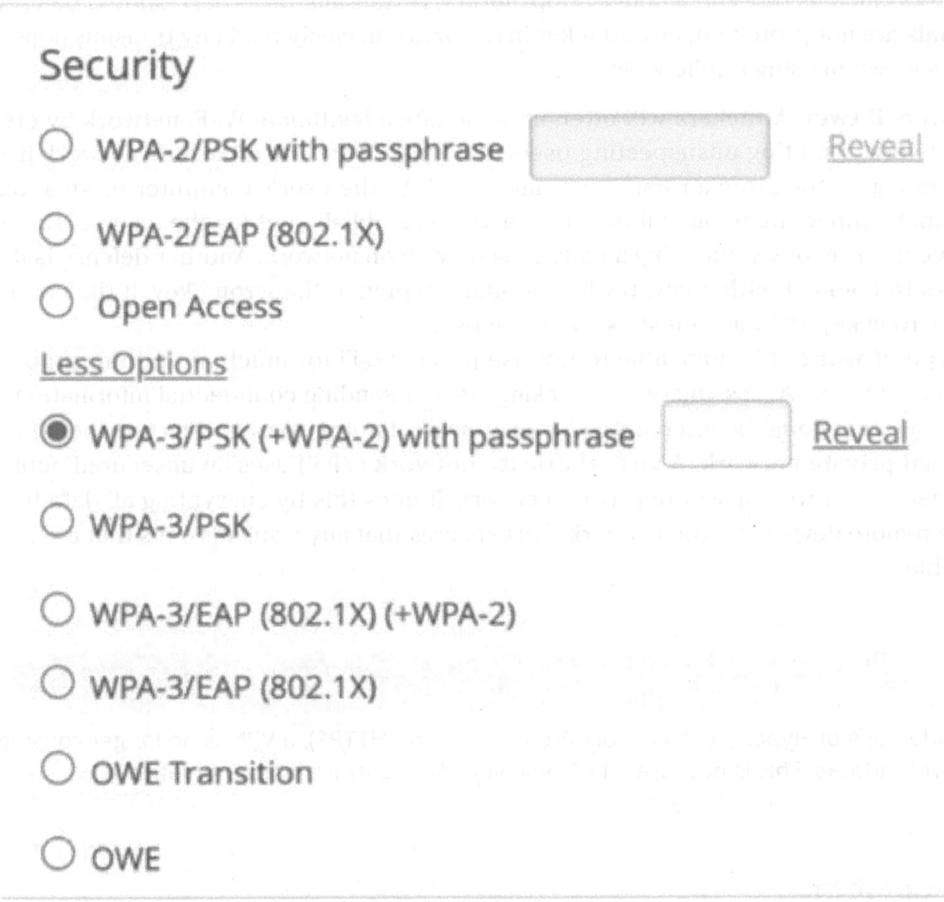

Other Security Settings Although securing the wireless router with a strong password and turning on WPA Personal are the most effective Wi-Fi security settings, other security settings can add another degree of security:

- **Change the SSID**. The name of the wireless network can be set by the user and is called the *Service Set Identifier (SSID)*. All wireless routers come with a default SSID. An attacker who picks up a Wi-Fi signal and can read the SSID could determine the type of wireless router being used and can exploit any weaknesses of that type of router. The SSID on the wireless router should be changed from its default value to an anonymous value that does not identify the owner or location of the network. For example, *SULLIVAN_HOUSE* or *1234_Main_St* would not be good SSIDs; a better choice might be something like *MyWireNet599342*.
- **Turn on guest access**. Most wireless routers allow for a separate guest network to be set up in addition to the main Wi-Fi network. This serves to isolate the main network from the guest network. The guest network can be configured so that any user who connects to the separate guest network can only access the Internet directly and other devices in the same network. Another option restricts guests to only Internet access; they cannot access any other network devices, such as a printer.

Exam Tip ✔

Although widely advertised, other security settings only provide a slight degree of additional security or no additional security at all. These include *disabling SSID broadcasts, restricting users by MAC address*, and *limiting the number of users*. It is not recommended that these settings be used.

Using Public Wi-Fi

Public Wi-Fi networks, such as those in a coffee shop, library, restaurant, or airport, should be used with a degree of caution. If the signals are not protected, any attacker in the area can easily read any transmissions. The following is a list of sound practices when using public Wi-Fi:

- **Watch for an evil twin**. Attackers will often impersonate a legitimate Wi-Fi network by creating their own lookalike network, tempting unsuspecting users to connect with the attacker's network instead. When the connection is made, the attacker can inject malware into the user's computer or steal data from it. One defense against connecting to an evil twin is to ask the establishment for the name of the official Wi-Fi network to prevent erroneously choosing an advertised evil twin network. Another defense is if the official Wi-Fi network asks to connect with a key, try intentionally typing in the wrong key. If the connection blatantly accepts the wrong key, then it is most likely an evil twin.
- **Limit the type of work**. It is advisable to not use public Wi-Fi for much more than simple web surfing or watching online videos. Accessing online banking sites or sending confidential information such as a Social Security number that could be intercepted by an attacker if not properly protected should be avoided.
- **Using a virtual private network**. A **virtual private network (VPN)** uses an unsecured public network, such as the Internet, as if it were a secure private network. It does this by encrypting all data that is transmitted between the remote device and the network. This ensures that any transmissions that are intercepted will be indecipherable.

Note 19

Due to the expanded use of Hypertext Transport Protocol Secure (HTTPS), a VPN is no longer considered an essential defense against Wi-Fi attacks. This is because HTTPS encrypts the transmissions automatically.

Configuring Bluetooth

When using a smartphone or tablet that supports Bluetooth, it is advisable to disable Bluetooth and turn on this service only as necessary. To prevent bluesnarfing, Bluetooth devices should be turned off when not being used or when in a room with unknown people. Another option is to set Bluetooth on the device as *undiscoverable*, which keeps Bluetooth turned on in a state where it cannot be detected by another device.

Mobile Device Security

Securing mobile devices requires several steps. These include setting the cybersecurity configurations of the device, following best practices, and dealing with the theft or loss of the device.

Device Configuration

Several security configurations should be considered for a mobile device, including disabling unused features and enabling screen locks.

Disable Unused Features Mobile devices have a wide variety of features for the user's convenience. However, each of these can also serve as a threat vector. It is important to disable unused features and turn off those that do not support the business use of the phone or that are rarely used. To prevent bluejacking and bluesnarfing, one feature that should be disabled if it is not being regularly used is Bluetooth wireless data communication.

Enable Lock Screen A **lock screen** prevents the mobile device from being used until the user enters the correct passcode. Most mobile devices have different options for the type of passcode that can be entered. The most secure option is to configure a strong alphanumeric password on the mobile device. Other options include using biometrics such as facial recognition or fingerprint scan, or swiping a specific pattern of connecting dots, as illustrated in Figure 5-11.

Figure 5-11 Swipe pattern

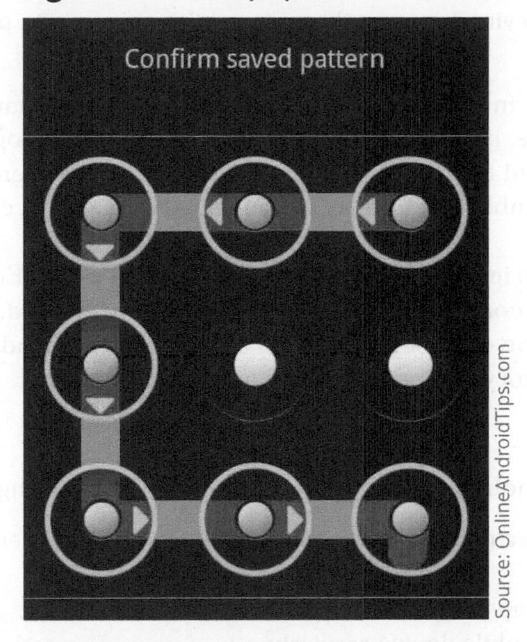

The least effective method is a short personal identification number or PIN. Many users opt to set a short four-digit PIN, similar to those used with a bank's automated teller machine (ATM). However, short PIN codes provide only a limited amount of security. An analysis of 3.4 million users' four-digit (0000–9999) PINs that were compromised revealed that users create predictable PIN patterns. The PIN *1234* was used in more than one out of every 10 PINs. Table 5-5 lists the five most common PINs and their frequency of use. Of the 10,000 potential PIN combinations, 26.83 percent of all PINs could be guessed by attempting just the top 20 most frequent PINs.[7]

Table 5-5 Most common PINs

PIN	Frequency of use
1234	10.71%
1111	6.01%
0000	1.88%
1212	1.19%
7777	0.74%

Note 20

The research also revealed that the least common PIN was *8068*, which appeared in only 25 of the 3.4 million PINs.

Lock screens should be configured so that the device must be unlocked before it can be used. Also, the device should be configured so that when it is left idle for some time, the device locks. Most mobile devices can be set to have the screen automatically lock after anywhere from 30 seconds to 30 minutes of inactivity.

Note 21

Unless at home, users must *always* lock the device *before* they set it down. Do not rely on the inactivity setting: a thief can easily grab the device and access it before the automatic lock takes effect.

The mobile device must also be configured so that after a specific number of failed attempts to enter the correct passcode, such as when a thief is trying to guess the code, additional security protections will occur. These include the following:

- **Extend lockout period**. If an incorrect passcode is entered a specific number of times, the lockout period will be extended. For example, if the incorrect passcode is entered five consecutive times, the mobile device will remain completely locked for one minute. If the incorrect code is entered again after one minute, the device will stay locked for double that time, or two minutes. For each successive incorrect entry, the lockout period will double.
- **Reset to factory settings**. If an incorrect passcode is entered a set number of times, the user will be prompted to enter a special phrase to continue. If the phrase is correctly entered, then the user will have only one more opportunity to enter the correct passcode. If an incorrect passcode is entered again, the device will automatically reset to its factory settings and erase any data stored on it.

Best Practices

A list of "best practices" for using mobile devices securely includes the following:

- Never jailbreak or root a smartphone. This also disables the built-in operating system security features on the phone.
- Do not sideload unapproved apps.
- Back up data stored on the mobile device regularly.
- Use appropriate sanitization and disposal procedures for mobile devices. Users should delete all information stored in a mobile device before discarding, exchanging, or donating it.
- Treat text messages the same as phishing emails. Do not call telephone numbers contained in unsolicited text messages or click links.

Device Loss or Theft

One of the greatest risks of a mobile device is the loss or theft of the device. Unprotected devices can be used to access web resources or view sensitive data stored on the device. Do the following to reduce the risk of theft or loss:

- Keep the mobile device out of sight when traveling in a high-risk area.
- Avoid becoming distracted by the device. Always maintain an awareness of your surroundings.
- When holding a device, use both hands to make it more difficult for a thief to snatch.
- Do not use the device on escalators or near transit train doors.
- White or red headphone cords may indicate they are connected to an expensive device. Consider using wireless earbuds instead.

If theft of a smartphone does occur, follow these steps in order:

1. **Call the phone or use the carrier's mobile app to send an alert**. Use another device to call the mobile phone. Even if the phone was on silent mode if it is on a hard surface, you may be able to hear it vibrate and locate it. You can also use your carrier's mobile app to override your ringer settings and sound an alert.
2. **Text the phone**. If you have text messages set to display on the lock screen, this can help a good Samaritan return your phone to you quickly. In the text, provide instructions on how to reach you in case someone finds the phone.
3. **Use the phone's built-in "find my phone" feature**. Modern phones have built-in security measures that allow you to track, ring, lock, or erase them remotely. Be sure that you have the feature enabled.
4. **Remotely erase data from the phone**. If you are fairly certain your smartphone was stolen, you should remotely delete the data from your device as soon as possible. Sophisticated thieves can make it impossible to wipe the device if you delay.

Note 22

In some instances, employees whose phone is used for business purposes or to receive company emails must agree in advance that a "remote wipe" will be performed if a phone is missing or there is a suspicion of theft or misuse of employer information.

5. **Lock the phone and change passwords**. If your cell phone cannot be located, use the phone's "find my phone" feature to remotely lock the phone. Change the passwords on any accounts you regularly access on the phone or for which you had an app installed on the phone, particularly financial accounts and email.
6. **Contact the mobile carrier**. If you cannot locate the device after erasing and locking it remotely, contact your cellular carrier. They can disable service to your phone and mark the phone itself as unusable, even if a new internal card is installed or someone attempts to purchase service for the phone at a different carrier.
7. **Alert the police**. You could potentially need a police report if you have to protest fraudulent charges made with your device.

Two Rights & A Wrong

1. The first step in securing a wireless router is to create a strong password to protect its internal configuration settings.
2. There is no known defense against connecting to an evil twin.
3. To prevent bluesnarfing, Bluetooth devices should be turned off when not being used or when in a room with unknown people.

See Appendix A for the answer.

Module Summary

- Wi-Fi is a wireless data network technology that provides high-speed data connections for mobile devices. Each mobile device must have a wireless adapter to send and receive the wireless signals, and these adapters are built into mobile devices. In addition, a wireless router is needed for a home-based Wi-Fi network. The router acts as the "base station" for wireless devices, sending and receiving wireless signals between all devices as well as providing the "gateway" to the external Internet. In an office setting, instead of using a wireless router, a more sophisticated device known as an access point (AP) is used. Several attacks can be launched against home Wi-Fi networks, such as stealing data, reading wireless transmissions, injecting malware, and downloading harmful content.
- Bluetooth technology is a short-range wireless technology designed for interconnecting two devices. Two Bluetooth attacks are bluejacking and bluesnarfing. Bluejacking is an attack that sends unsolicited messages to Bluetooth-enabled devices. Bluesnarfing is an attack that accesses unauthorized information from a wireless device through a Bluetooth connection, often between cell phones and laptop computers.
- Near field communication (NFC) is a set of standards used to establish communication between devices in very close proximity. When the devices are brought within 4 centimeters of each other or tapped together, two-way communication is established. Consumer NFC devices are often used as an alternative to cash or credit card payment methods, which is called a contactless payment system. NFC vulnerabilities include eavesdropping, data theft, man-in-the-middle attacks, and device theft.

- There are different types of mobile devices. Tablets are portable computing devices that are generally larger than smartphones and smaller than laptops and are focused on ease of use. A smartphone has an operating system that allows it to run apps and access the Internet. A popular class of mobile technology consists of devices that can be worn by the user instead of carried. Known as wearable technology, these devices can provide even greater flexibility and mobility than smartphones. As a class, portable computers are devices that closely resemble standard desktop computers but are smaller self-contained devices that can easily be transported from one location to another while operating on battery power. A popular type of computing device that resembles a laptop computer is a web-based computer. It contains a limited version of an OS and a web browser. It is designed to be used while connected to the Internet so that no traditional software applications can be installed, and no user files are stored locally on the device.
- Several risks are associated with using mobile devices. However, users can circumvent the installed built-in limitations on their smartphone (called jailbreaking on Apple iOS devices or rooting on Android devices) to download from an unofficial third-party app store (called sideloading). Because these apps have not been vetted, they may contain security vulnerabilities or malicious code. Another means by which untrusted content can infect a mobile device is through text messages. Mobile devices can also access untrusted content that other types of computing devices generally cannot. One example is Quick Response (QR) codes, through which attackers can direct the web browser to the attacker's imposter website or to a site that immediately downloads malware.
- Portability is the greatest asset of a mobile device. It is also one of its greatest vulnerabilities: mobile devices are routinely lost or stolen. Security patches and updates for Android and macOS are distributed through over-the-air (OTA) updates. Apple commits to providing OTA updates for up to eight years after the OS is released. However, OTA updates for Android OSs vary considerably. Users rely on public Wi-Fi for Internet access. Because these networks are beyond the control of the user, the type of security that is available may be weak. In a public Wi-Fi network, an attacker can set up an evil twin to mimic an authorized Wi-Fi device. A user's mobile device may unknowingly connect to this evil twin instead of the authorized device so that attackers can receive any transmissions or directly send malware to the user's computer. Mobile devices using geolocation are at increased risk of targeted physical attacks. An attacker can determine where the user with the mobile device is currently located and use that information to follow the user to steal the mobile device or inflict harm upon the person.
- Reducing the risk of attack through wireless networks is an important security step. The first step in securing a wireless router is to create a strong password to protect its internal configuration settings. Wireless routers may not regularly receive important patches from their OEMs. If a wireless router does not receive regular updates, it may be necessary to search the wireless router configuration settings or OEM website to determine if there is a patch to update the router. Many routers also permit remote management of the router's configuration settings through the Internet. Turning remote management off adds a stronger degree of security because it limits access to the configuration settings of the wireless router to only the local computer connected to it.
- For all devices other than Wi-Fi 6E devices, Wi-Fi Protected Access 2 (WPA2) Personal provides a high level of wireless security. Implementing WPA2 Personal involves turning it on at the wireless router and then entering a key value on each authorized device that has been preapproved to join the Wi-Fi network. For devices based on Wi-Fi 6E, Wi-Fi Protected Access 3 (WPA3) Personal provides the optimum level of wireless security. Public Wi-Fi networks should be used with a degree of caution. Users should watch for an evil twin, limit the type of work done on a public Wi-Fi network, and if necessary, use a virtual private network (VPN). When using a smartphone or tablet that supports Bluetooth, it is advisable to disable Bluetooth and turn on this service only as necessary.
- Mobile devices include a wide variety of features for the user's convenience. It is important to disable unused features and turn off those that do not support the business use of the phone or that are rarely used. A lock screen prevents the mobile device from being used until the user enters the correct passcode. Most mobile devices have different options for the type of passcode that can be entered. Lock screens should be configured so that the device must be unlocked before it can be used. Also, the device should be configured so that when it is left idle for some time, the device will lock. Several "best practices" should be followed for securing a mobile device. If a device is lost or stolen, a series of important steps should be taken.

Key Terms

access point (AP)
bluejacking
bluesnarfing
Bluetooth
contactless payment system
evil twin
geolocation
GPS tagging (*geotagging*)
Institute of Electrical and Electronics Engineers (IEEE)
jailbreaking
lock screen
mesh
multimedia messaging service (MMS)
near field communication (NFC)
portable computer
Quick Response (QR)
rich communication services (RCS)
rooting
short message service (SMS)
sideloading
smartphone
tablet
virtual private network (VPN)
wearable
Wi-Fi
Wi-Fi Protected Access 2 (WPA2) Personal
Wi-Fi Protected Access 3 (WPA3) Personal
Wi-Fi Protected Setup (WPS)
wireless adapter
wireless router

Review Questions

1. Alyona has been asked to research a new payment system for the retail stores that her company owns. Which technology is predominately used for contactless payment systems that she will investigate?
   a. Wi-Fi
   b. Bluetooth
   c. Near field communication (NFC)
   d. RF-FA
2. Taisiya is investigating a security incident in which confidential data was stolen from the CFO's smartphone. As she examines the phone, she notes that Bluetooth is turned on, which is a violation of policy for company phones. She suspects that it was an attack that used Bluetooth. Which attack would this be?
   a. Blueswiping
   b. Bluesnarfing
   c. Bluejacking
   d. Bluestealing
3. Which of these is NOT a risk when a home wireless router is not securely configured?
   a. User names, passwords, credit card numbers, and other information sent over the Wi-Fi could be captured by an attacker.
   b. Malware can be injected into a computer connected to a Wi-Fi network.
   c. Patches for Wi-Fi wireless routers could not be applied.
   d. An attacker can steal data from any folder with file sharing enabled.
4. Which of these Wi-Fi Protected Setup (WPS) methods is vulnerable?
   a. Click-to-connect method
   b. PIN method
   c. Net method
   d. Push-button method
5. Olga has been asked to provide research regarding adding a new class of Android smartphones to a list of approved devices. One of the considerations is how frequently the smartphones receive firmware OTA updates. Which of the following reasons would Olga NOT list in her report as a factor in the frequency of Android firmware OTA updates?
   a. OEMs are hesitant to distribute Google updates because it limits their ability to differentiate themselves from competitors if all versions of Android start to look the same through updates.
   b. Because many of the OEMs had modified Android, they are reluctant to distribute updates that could potentially conflict with their changes.
   c. Wireless carriers are reluctant to provide firmware OTA updates because of the bandwidth it consumes on their wireless networks.
   d. Because OEMs want to sell as many devices as possible, they have no financial incentive to update mobile devices that users would then continue to use indefinitely.

6. **What is the process of identifying the geographical location of a mobile device?**
  a. Geotracking
  b. Geolocation
  c. GeoID
  d. Geomonitoring

7. **Which of these could be used by an attacker to send a specially crafted video that can introduce malware into a device via a text message?**
  a. MMS
  b. RSR
  c. CRS
  d. SRS

8. **Arina's nephew downloaded and installed an app that allows him to circumvent the built-in limitations on his Android smartphone. What is this called?**
  a. Ducking
  b. Sideloading
  c. Jailbreaking
  d. Rooting

9. **Which of the following prevents a mobile device from being used until the user enters the correct passcode?**
  a. Touch swipe
  b. Swipe code
  c. Timeout
  d. Screen lock

10. **Which of these is considered the strongest type of passcode to use on a mobile device?**
  a. Password
  b. PIN
  c. Fingerprint swipe
  d. Draw connecting dots pattern

11. **What is the technical name for a Wi-Fi network?**
  a. Wireless local area network (WLAN)
  b. Wireless personal area network (WPAN)
  c. Wireless ultraband
  d. Bluetooth

12. **What is the strength of a tablet computer?**
  a. Processing capabilities
  b. Ease of use
  c. Wireless connection speed
  d. Hardware upgrades

13. **Which of the following is FALSE about a wireless router?**
  a. It sends and receives wireless signals between all wireless devices.
  b. It is usually found in a large business with thousands of wireless users.
  c. It typically is connected to the user's modem.
  d. It combines several networking technologies.

14. **What is it called when a user moves from one area of coverage to another area of coverage in a Wi-Fi network?**
  a. Migrating
  b. Traveling
  c. Pushing
  d. Roaming

15. **Why are Android devices considered to be at a higher security risk than Apple iOS devices?**
  a. Android apps can be sideloaded.
  b. Apple apps are written in a more secure binary language.
  c. iOS is a simpler operating system than Android.
  d. Android apps are written in JavaScript.

16. **What is the first step in securing a Wi-Fi wireless router?**
  a. Monitor the RF signal with a remote telemonitor.
  b. Turn on short preamble packets.
  c. Disable all wireless connections.
  d. Create a password to protect it.

17. **Which of the following provides the optimal level of wireless security for a home Wi-Fi network?**
  a. WPA1
  b. WPA3
  c. WPA5
  d. WPA9

18. **Which of the following is used by attackers to attempt to capture the transmission from legitimate users?**
  a. Evil parent
  b. Evil sister
  c. Evil brother
  d. Evil twin

19. **Which of the following is NOT considered a best practice for using mobile devices securely?**
  a. Treat text messages the same as phishing emails.
  b. Use appropriate sanitization and disposal procedures.
  c. Limit the number of backups.
  d. Avoid sideloading apps.

20. **Lyubov discovered that her smartphone is missing. What is the recommended first step for her to take?**
  a. Contact the police.
  b. Call the phone.
  c. Remotely erase data from the phone.
  d. Contact the mobile carrier.

Hands-On Projects

Project 5-1: Using Microsoft Windows Network Shell Commands

Microsoft Windows displays the network status of a device that is connected to a network, either a wired network or a Wi-Fi network. However, configuring different connection parameters is largely unavailable to users through the main interface. However, the Windows Network Shell (*netsh*) can be used to configure and display advanced network settings. Netsh is a *command-line interface* (*CLI*) that requires commands to be typed instead of selected from a menu. This gives netsh extended flexibility. In this project, you will explore some of the *netsh* commands.

Note 23

For this project, you will need a computer running Microsoft Windows that has a wireless NIC and can access a Wi-Fi network.

1. In Microsoft Windows, right-click the **Start** button.
2. Select **Windows Terminal (Admin)**. This will open the Windows command window in elevated privilege mode.
3. Type **netsh** and then press **Enter**. The command prompt will change to *netsh*>.
4. Type **wlan** (an abbreviation for *wireless local area network*) and then press **Enter**. The command prompt will change to *netsh wlan*>.
5. Type **show drivers** and then press **Enter** to display the wireless adapter information. It may be necessary to scroll up to see all the information.
6. Under **Radio types supported**, the different types of Wi-Fi networks to which this computer can connect are displayed. What is the maximum data rate of the highest level of radio type?

Note 24

The "radio types supported" are listed by their IEEE nomenclature. Refer to Table 5-1 for a listing of these and the corresponding Wi-Fi Alliance versions.

7. Under **Authentication and cipher supported in infrastructure mode**, which Wi-Fi Protected Access Personal is supported, WPA2 Personal or WPA3 Personal? Why is this version supported?
8. View the WLAN interfaces for this computer. Type **show interfaces** and then press **Enter**. Record the SSID value and the name of the Profile.
9. Display all the available Wi-Fi networks to this computer. Type **show networks** and then press **Enter**.
10. Windows creates a profile for each network to which you connect. To display those profiles, type **show profiles** and then press **Enter**. If there is a profile of a network that you no longer use, type **delete profile name=***profile-name.*
11. Disconnect from your current WLAN by typing **disconnect** and then press **Enter**. Note the message you receive and observe the status in your system tray.
12. Reconnect to your network by typing **connect name=***profile-name* **ssid=***ssid-name* as previously recorded and then press **Enter**.
13. Netsh allows you to block specific networks. Select another network name that you currently are not connected to. Type **show networks**, press **Enter**, and then record the SSID of that network you want to block.
14. Type **add filter permission = block ssid=***ssid-name* **networktype = infrastructure** and then press **Enter**.

Continued

15. Type **show networks** and then press **Enter**. Does the network that you previously blocked appear in the list?
16. Display the blocked network (but do not allow access to it). Type **set blockednetworks display=show** and then press **Enter.**
17. Type **show networks** and then press **Enter**. Does the network that you previously blocked appear in the list?
18. Click the wireless icon in your system tray. Does the network appear in this list?
19. Click the wireless icon in your system tray again. What appears next to the name of this blocked network?
20. Now re-enable access to the blocked network by typing **delete filter permission = block ssid=***ssid-name* **networktype = infrastructure** and then press **Enter**.
21. Netsh can also generate a detailed Wi-Fi network report. This report shows all the Wi-Fi events from the last three days and groups them by Wi-Fi connection sessions. It also shows the results of several network-related command line scripts and a list of all the wireless adapters. Type **show wlanreport** and then press **Enter**.
22. Record the location in which the report was written.
23. Open a web browser and type the location of the report. For example, if the location is *C:\ProgramData\Microsoft\Windows\WlanReport\wlan-report-latest.html* then enter into the web browser **file:/// C:\ProgramData\Microsoft\Windows\WlanReport\wlan-report-latest.html**
24. Scan the report for the information. Note that the first paragraph displayed allows you to hover over a session or click it. How could this information be helpful?
25. Type **Exit** and then press **Enter**.
26. Type **Exit** again and then press **Enter** to close the command window.

Project 5-2: Using a Wireless Monitor Tool

Most Wi-Fi users are surprised to see just how far their wireless signal will reach, and if the network is unprotected, this makes it easy for an attacker hiding several hundred feet away to break into the network. Several tools are available to show the different wireless signals from Wi-Fi networks that can be detected. In this project, you download and install the NirSoft WifiInfoView tool. You will need a computer with a wireless adapter, such as a laptop, to complete this project.

1. Use your web browser to go to **https://www.nirsoft.net/utils/wifi_information_view.html**. (If you are no longer able to access the site through the web address, use a search engine to search for "NirSoft WifiInfoView.")
2. Scroll down and click **Download WiFiInfoView (32-bit)** or **Download WiFiInfoView (64-bit)** depending on your system.
3. Download the tool and when finished, extract the files and then launch the program.
4. Wait until WiFiInfoView displays all Wi-Fi networks that it detects. If necessary, maximize the WiFiInfoView window.
5. Scan all the information that is displayed for each Wi-Fi network. Does the amount of available information from Wi-Fi networks to which you have not connected surprise you?
6. Scroll back to the first column of information.
7. Under **SSID**, does each network have a service set identifier? Why would an SSID not appear? Does disabling the broadcast of the SSID name give an enhanced level of security? Why not?
8. Under **RSSI**, find the signal strength (lower numbers indicate a stronger signal).
9. Note the value in the **Signal Quality** column. How is the "strength" of the signal different from the "quality" of the signal?
10. Note the information provided under **Company**, **Router Model**, and **Router Name**. Since this information is broadcast to anyone who has this program, how could an attacker use this information in forming an attack?
11. Under **Security**, are any Wi-Fi networks listed that are not WPA2 or WPA3? How could an attacker use this information?
12. Under **WPS Support**, are there networks that do not support WPS? Why is this a good security setting?
13. Note the number listed under **Stations Count**. How could this number be used for security?

14. Double-click the Wi-Fi network to which you are currently connected. A window of the available information that is being transmitted through the Wi-Fi is displayed. This is the information that anyone can see regarding your Wi-Fi network. Close this window.
15. Now select a network other than the one to which you are connected and double-click it to display information. This is the information that anyone can see regarding this Wi-Fi network to which you are not connected. After reading the information, close the window.
16. What additional information do you find useful? What information would a threat actor find useful?
17. Close all windows.

Project 5-3: Creating and Using QR Codes

Quick Response (QR) codes can be read by an imaging device such as a mobile device's camera or online. However, they pose a security risk. In this project, you create and use QR codes.

1. Use your web browser to go to **www.qrstuff.com**. (If you are no longer able to access the program through this URL, use a search engine to search for "Qrstuff.")
2. First, create a QR code. Under **DATA TYPE**, be sure that **WEBSITE URL** is selected.
3. Under **CONTENT**, enter the URL **http://www.cengage.com**. Watch how the **QR CODE PREVIEW** changes as you type.
4. Under **Encoding Options**, select **Static - Embed URL into code as-is**.
5. Click **DOWNLOAD QR CODE** to download an image of the QR code.
6. Navigate to the location of the download and open the image. Can you tell anything by looking at this code? How could a threat actor use this to his advantage? Where could malicious QR codes be used? Does a user have any protection when using QR codes?
7. Now use an online reader to interpret the QR code. Use your web browser to go to **blog.qr4.nl/Online-QR-Code-Decoder.aspx**. (The location of the content on the Internet may change without warning. If you are no longer able to access the program through this URL, use a search engine and search for "Free Online QR Code Reader.")
8. Click **Choose File**.
9. Navigate to the location of the QR code that you downloaded on your computer and click **Open**.
10. Click **Upload**.
11. In the text box, what is displayed? How could an attacker use a QR code to direct a victim to a malicious website?
12. Use your web browser to go to **www.qrcode-monkey.com**. (If you are no longer able to access the program through this URL, use a search engine to search for "QRcodemonkey.")
13. Click **LOCATION**.
14. On the map, drag the pointer to an address with which you are familiar. Note how the **Latitude** and **Longitude** change.
15. Click **Create QR Code**.
16. Click **Download PNG** to download this QR code to your computer.
17. Navigate to the location of the download and open the image. How does it look different from the previous QR code? Is there anything you can tell by looking at this code?
18. Use your web browser to return to **https://blog.qr4.nl/Online-QR-Code-Decoder.aspx**.
19. Click **Choose File**.
20. Navigate to the location of the map QR code that you downloaded on your computer and click **Open**.
21. Click **Upload**.
22. In the text box, a URL will be displayed. Paste this URL into a web browser.
23. What does the browser display? How could an attacker use this for a malicious attack?
24. Return to **www.qrstuff.com**.
25. Click each option under **DATA TYPE** to view the different items that can be created by a QR code. Select three and indicate how they could be used by an attacker.
26. Close all windows.

Continued

Project 5-4: Configuring a Wireless Router

The ability to properly configure a wireless router is an important skill for end-users. In this project, you use an online emulator from TRENDnet to configure a wireless router.

1. Use your web browser to go to **https://www.trendnet.com/emulators/TEW-827DRU_v2.0R/basic_status.html**. (The location of the content on the Internet may change without warning; if you are no longer able to access the program through this URL, use a search engine and search for "Trendnet Emulators.")
2. An emulated Setup screen displaying what a user would see when configuring an actual TRENDnet router is displayed.
3. Be sure that the **BASIC** tab is selected in the left pane. Note the simulated **Network Status** information.
4. Click **Wireless** in the left pane and read the information displayed.
5. Under **Broadcast Network Name (SSID)**, note the two options. What would **Disable** do? Why is this not considered a strong security step (consider Step 7 of Hands-On Project 5-2)?
6. Under **Security**, note the default setting for **Security Mode**. Is this a good option default option?
7. Under **Security**, use the pull-down menu to display the options for **Security Mode**. What does **WPA2-PSK** mean?
8. Click the down arrow on **WPA2-PSK**. What are the other options?
9. Under **WPA**, the **Pre-Shared Key** is the value that would be entered on this wireless router and on each of the wireless devices on the network. Click **Show Password** and then enter a strong key value.
10. In the left pane, click **Guest Network**. Read the information about a guest network. A guest network allows you to have an additional open network just for occasional guests that does not affect the main wireless network. How could this be an advantage?
11. Note the option under **Internet Access Only**. When would you select this option?
12. Note the option under **WLAN Partition**. Why is this not enabled by default?
13. Under **Security**, note that an option under **Security Mode** is **Disable**. Why would a guest network's security be turned off by default? (Hint: If it were turned on, what would the guests need before they could use the network?)
14. In the left pane, click **Advanced**.
15. Click **Security**.
16. Under **Access Control**, what is the **Enable Access Control**? Does it provide strong security if it were enabled?
17. Click **Setup**.
18. Click **Upload Firmware** and read through the information. When would you use this option?
19. How easy is this user interface to navigate? Does it provide enough information for a user to set up the security settings on this system?
20. Close all windows.

Case Projects

Case Project 5-1: Your Wi-Fi Security

Is the wireless network you own as secure as it should be? Examine your wireless network or that of a friend or neighbor and determine its level of security (if necessary, use the tool in Hands-On Project 5-2). Next, outline the steps it would take to make it stronger. Estimate how much it would cost and how much time it would take to increase the level. Finally, estimate how long it would take you to replace all the data on your computer if it was corrupted by an attacker, and what you might lose. Would this be motivation to increase the security? Write a one-page paper on your work.

Case Project 5-2: Compare Wireless Routers

Use the Internet to identify three different brands of wireless routers. Create a table that lists each device, its features, and costs. Then research information about the security of each wireless router. Check the supporting website for information about patches. Below the table, write a paragraph describing which you would choose for your home use and why.

Case Project 5-3: Survey of Wireless Users

Create a short survey to administer to wireless users regarding how they would typically use a free wireless network in a restaurant or coffee shop. Include questions such as, "What precautions do you take when using a free wireless network?" "Can you list the dangers of using an open wireless network?" "Do you ever purchase anything that requires you to type in a credit card number while using a free wireless network?" and "Do you know how to spot an evil twin?" Ask five friends or acquaintances for their responses. Based on these responses, give a grade of "A" through "F" to each wireless user. Now take the test yourself and give yourself a similar rating. What improvements can you make in using a free wireless network?

Case Project 5-4: Attacks on Wireless Medical Devices

Many medical devices use wireless technology, yet they lack the necessary security protections. At a security conference, a security researcher, who was himself a diabetic, demonstrated a wireless attack on an insulin pump that could change the delivery of insulin to the patient. A security vendor found that they could scan a public space from up to 300 feet (91 meters) away, find vulnerable pumps made by a specific medical device manufacturer, and then force these devices to dispense fatal insulin doses. And another researcher "hacked" into a defibrillator used to stabilize a heartbeat and reprogrammed it. He also disabled its power-save mode, so the battery ran down in hours instead of years. Use the Internet to research the current state of attacks on wireless medical devices and proposed defenses. Should the vendors who make these wireless medical devices be forced to add security features to their devices? What should be the penalty if they do not? What should be the penalty for an attacker who manipulates a wireless medical device? Write a one-page paper on your research.

Case Project 5-5: Bluetooth Range Estimator

The range at which a Bluetooth device can transmit depends on several different factors. It is good to have an understanding of ranges to be aware of whether or not a Bluetooth-enabled device could be the victim of a Bluejacking or Bluesnarfing attack. Go to **https://www.bluetooth.com/learn-about-bluetooth/bluetooth-technology/range/** to explore the Bluetooth Range Estimator tool. First, watch the video and then read the details of each of the key factors. Then use the range estimator tool, changing the different parameters (receiver sensitivity, path loss, transmit power, transmitter antenna gain, and receiver antenna gain) to determine the estimated range. What does this tell you about Bluetooth ranges? How could this tool be used? Write a one-page paper on what you have learned.

References

1. Clement, J., "Share of global mobile website traffic 2015-2021," *Statista*, Feb. 18, 2022, accessed May 13, 2022, https://www.statista.com/statistics/277125/share-of-website-traffic-coming-from-mobile-devices/.
2. Wise, Jason, "40 Smartphone statistics 2022: How many people have smartphones?" *Earthweb*, May 12, 2022, accessed May 13, 2022, https://earthweb.com/smartphone-statistics/.
3. "What to do if your phone is lost or stolen," *Asurion*, accessed May 14, 2022, https://www.asurion.com/connect/tech-tips/what-to-do-when-your-phone-is-lost-or-stolen/.
4. "Laptop and mobile device theft awareness," *University of Pittsburgh*, accessed May 27, 2020, https://www.technology.pitt.edu/security/laptop-theft.
5. "Data gets personal: 2019 global data risk report from the Varonis data lab," *Veronis*, accessed May 27, 2020, https://www.varonis.com/2019-data-risk-report/.
6. Wagenseil, Paul, "The one router setting everyone should change (but no one does)," *Tom's Guide*, Apr. 13, 2018, accessed May 14, 2022, https://www.tomsguide.com/us/change-router-default-passwords,news-26975.html.
7. "Pin analysis," *DataGenetics*, accessed Mar. 10, 2014, http://datagenetics.com/blog/september32012/index.html.

Module 6

Privacy

After completing this module, you should be able to do the following:

1. Explain how data is being stolen from users.
2. Identify "data thieves."
3. Describe the risks associated with data theft.
4. Define cryptography and explain how it can provide protection.
5. Explain how to strengthen privacy through limiting cookies, disabling MAIDs, and following privacy best practices.

Cybersecurity Headlines

The COVID-19 global pandemic that swept around the world in 2020 had a colossal impact on every country. Governments, which were unprepared for this unprecedented event, struggled to find the best solution to limit the spread of the virus. Some governments, such as the United States and United Kingdom, delayed stringent lockdowns. Other nations acted more quickly and aggressively: they closed borders, enacted travel restrictions, suspended schooling, and canceled all public events. They also performed meticulous testing and "contact tracing" (warning those who came in contact with an infected individual) to attempt to break the chain of transmission.

One nation that stood out for its aggressive response to COVID-19 was the tiny central African nation of Rwanda. The most densely populated nation on the African mainland where 13 million people live in an area roughly the size of Maryland, Rwanda was praised as a "poster child" for how to tackle COVID-19 on the continent. Six months after the initial outbreak of the virus, Rwanda only registered 4,800 cases and 29 deaths, a tiny fraction of what other countries were experiencing. It became only one of 11 countries worldwide, and the only country in Africa, that the European Union (EU) at that time declared a safe travel destination.

However, the containment strategy used by Rwanda also raised concerns among privacy advocates. These advocates claim that Rwanda was overly aggressive in its use of technology and procedures, which impinged on its citizen's privacy.

Drone technology played a significant role in Rwanda's methodology. The government used over 60 drones to ferry protective equipment, COVID-19 test samples, and intravenous fluids to and from hospitals. However, they also used drones with loudspeakers to broadcast virus-prevention guidelines while hovering above city suburbs.

At intersections in major cities, Rwandan health workers chose drivers at random to stop and provide nasal samples (by law the drivers were required to comply). The samples were sent to government labs, with results usually delivered within 24 hours. Anyone who tested positive was then put under contact tracing and ordered to report to a government-run COVID-19 clinic for a lengthy stay. And any friends, relatives, co-workers, or acquaintances who had been in contact with the infected person were also required to quarantine.

Continued

As many as 70,000 citizens were arrested for virus-related infractions during the six months after the restrictions were implemented. These infractions included not following social-distancing rules, violating night curfews, or failing to wear masks. Not wearing a mask normally carried a $26 fine on the first offense; however, after a second offense, the offender could be jailed for up to one year. Many of those arrested were detained and sent to large sports stadiums to remain under the watch of armed guards. While at the stadium, the arrestees were required to spend long nights listening to public-health virus-related messages broadcast over loudspeakers.

Rwanda's police force has stated that the technology deployed in the fight against COVID-19 will continue to be used in the future, but not to fight the virus. Instead, the technology will be used to increase its ability to maintain public order and control Rwanda's borders. As one police spokesman said, "Policing is more efficient with technology."

The basic goal of cybersecurity defenses is to prevent unauthorized individuals from accessing our data and devices. This includes preventing cybercriminals from performing ransomware attacks, insiders from unleashing logic bombs, and state actors from infecting computers with remote access Trojans (RATs).

However, beyond these malware and social engineering attacks from traditional threat actors, there are other unauthorized individuals who are accessing and using our data without our knowledge and permission. Yet these individuals are not classified as attackers and they do not break any laws, but they earn *hundreds of billions of dollars annually* using our data—without our permission or knowledge.

Who are these unauthorized individuals? How do they access our data? How do they use it to earn these huge sums of money? And what can we do to protect our data and keep it private?

In this module, you learn about privacy and what users can do to safeguard their data. You will first learn about this data theft, and then you will examine ways in which to limit the erosion of your privacy.

Data Theft

The topic of data theft involves knowing what is being stolen and how and identifying the data thieves. It also involves understanding the risks to users associated with the theft and usage of their data.

Caution !

"Stolen" in this context does not mean taking the data is illegal. Instead, stolen means it is gathered and used without the clear knowledge and express permission of the data owner.

What Is Being Stolen and How?

Understanding data theft means knowing the types of data that is taken. It also requires understanding how it is being stolen.

Data Types

One day, Rollin planned to pick up his granddaughter Mia to spend an afternoon in the park together. Before leaving his house, Rollin used his *smartphone to check the weather forecast and traffic conditions*. After picking up Mia, on the drive to the park, Rollin let her use his tablet to play a game, during which time an *ad appeared for a scooter*. As Rollin and Mia walked through the park, they stopped beside a pond to take a selfie and then *used a filtering app* (to add a duck bill to Rollin) *before uploading it to his social media*. On the drive home. they *stopped at a yogurt store* for a treat.

Although this scenario seems like a risk-free day in the park, in reality, it is filled with numerous examples of data theft:

- **Smartphone to check the weather forecast and traffic conditions**. Tracking features (*trackers*) are embedded in virtually every app on a smartphone; in fact, the average app has *six* trackers. Trackers allow third parties to collect data from the user's interaction with the app. On Rollin's drive to the park, four different apps were collecting and tracking his location—even though these apps did not need to. (One of the apps was a flashlight app he recently downloaded.) All this data was then linked with all other tracker data to build a continuously updated profile on Rollin. This data is also sold to others.

Note 1

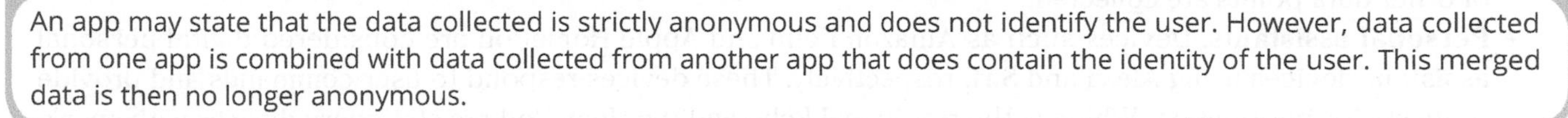

An app may state that the data collected is strictly anonymous and does not identify the user. However, data collected from one app is combined with data collected from another app that does contain the identity of the user. This merged data is then no longer anonymous.

- **Ad appeared for a scooter**. Although neither Rollin nor Mia had been looking online at scooters, nevertheless the scooter ad was no fluke. The scooter company had contracted with a *third-party ad network* to bid on the ad space for this tablet because they wanted to target people exactly like Rollin, who are living in the same city with his income bracket and have a young granddaughter. And this ad will follow Rollin and Mia for several days across all their different technology devices, appearing in multiple apps and on ads next to websites.

Note 2

Real-time bidding (*RTB*) is the process in which digital advertising is purchased. The entire RTB process is highly automated and occurs in less than one second.

- **Used a filtering app before uploading it to his social media**. The filtering app on Rollin's smartphone is not just limited to the selfie taken by the pond. Rather, it can access *any* of the photos on the smartphone along with data about the picture, such as its precise location. The social media app on Rollin's phone used to upload the selfie links the trip to the park with his current online activity along with data collected by other apps, such as his demographic information, purchasing habits, email address, zip code, and phone number.
- **Stopped at a yogurt store**. Because the trackers know where he stopped, and because Rollin paid for the yogurt with his credit card, this adds even more information to the comprehensive data profile of his preferences. The information is combined with data that shows he has a young granddaughter, which now results in Rollin's devices receiving targeted ads for yogurt and ice cream.

This is barely the tip of the iceberg of what data is being collected. Every time a user interacts with technology, they leave behind a "data trail," which is a digital record of their activity. This includes activities such as sending an email, browsing the Internet, using a smartphone app, making a purchase—or *any* activity using technology. And as more and more activities of a user's everyday life are performed electronically using more and more interconnected devices, the volume of data compiled on each user grows exponentially each day.

Consider just a few other examples:

- **Household appliances**. It is becoming increasingly difficult to purchase an appliance or even a smaller household "gadget" that does not require an Internet connection. Almost all high-end washing machines, dishwashers, dryers, refrigerators, and ovens require Wi-Fi connectivity. These connections allow the manufacturers to gather data on user activities and then transport the data back to their company—which it then combines with other data to build a comprehensive profile. Some companies are now intentionally disabling features until the appliance is connected to a Wi-Fi network—even though these features have been available on regular appliances *for over 75 years*.

- **Televisions**. Purchasing a "dumb" television is essentially impossible today. All TVs have Wi-Fi connections to support streaming and other features. But these TVs also actively collect user data on their viewing habits to determine which new shows to display or to advertise products that can be purchased directly from the TV. One television manufacturer posted an earnings report that said in one year the company lost $52 million from sales of TVs. However, this same company also identified "targeting using first-party data" that it collects from its customers as its fastest-growing revenue segment, offsetting the loss in TV sales.
- **Automobiles**. New vehicles have been called "computers with wheels." This is because today's cars gather *25 billion bytes (gigabytes) per hour of driving*, which is the equivalent of five high-definition movies. Information such as the vehicle's location, speed, braking, how specific components are operating, and any number of other data points are collected.
- **Personal assistants**. Devices such as Amazon Echo and Apple HomePod are considered digital personal assistant devices using Alexa and Siri, respectively. These devices respond to user commands and provide requests for information. Whereas the traditional Echo and the HomePod are stationary devices with speakers, new functionalities are continually being added. The Google Nest Hub Max contains a screen as well as speakers while Amazon Echo Show 10 is a robot that follows the user around the house. These devices continuously listen to conversations and gather data.
- **Web forms**. It is no surprise that information entered into a form on a webpage is captured and used for tracking. However, what may be surprising is the *point in time* when entered data is captured. Research has revealed that when entering a password or email address in a web form, the *Submit* button does not have to be clicked before the data is captured; instead, *as the data is being entered* by the user it is being captured. That means that typing a password or email address in a web form but then clicking *Cancel* does not stop the web form from capturing the data.

How Data Is Exfiltrated

Today the overwhelming majority of data that is stolen (*exfiltrated*) through trackers is from users' smartphones. These mobile devices have become the primary communications tool for accessing the Internet, sending text messages, taking videos and photos, and making phone calls. However, smartphones have two significant deficiencies. First, they were not designed to protect users' privacy. Second, smartphones give users much less control over the device than a standard computer. On a smartphone, it is almost impossible to install a different operating system, it is harder to remove or replace potentially unwanted programs (PUPs) that are installed by default, and it is very difficult to completely prevent tracking.

Note 3

PUPs are covered in Module 3.

Smartphone data exfiltration is primarily based on two smartphone features that are part of all smartphones today. These are location services and mobile advertising identifiers.

Location Services Virtually all modern smartphones support *geolocation*, or technology that can identify the geographical location of the device. In an open area that does not have tall buildings that can block a signal, geolocation is based on the Global Positioning System (GPS) satellite-based navigation system. However, in dense urban areas, Wi-Fi positioning is often used instead. The smartphone will scan the airwaves for Wi-Fi access points (APs) and measure the signal strength to each network. It then uses this data to estimate its own location.

Note 4

Geolocation, Wi-Fi, and APs are covered in Module 5.

As smartphones determine their current location, this information is then packaged into a feature on smartphones known as **location services.** Apps on the smartphone can request the current location data and then can use it to provide services that are based on location, such as displaying a traffic map of the current area. Many of these apps also transmit the location back to the company behind the app (called the *mother ship*) to track users and also share or sell the information to others.

Combining location services among multiple apps can result in a more detailed picture of the user. Location services have been used to determine a user's history of activities, beliefs, participation in events, and personal relationships. They have even been used to identify a journalist's confidential source.

Note 5

Some smartphones provide users a degree of control over whether apps have permission to use location services. However, not all apps may ask for permission, and research has shown that even some apps that are denied permission still continue to track users.

Mobile Advertising Identifier (MAID) In addition to location services, apps may share information about interactions the smartphone user has with apps, such as which apps have been installed, when they are opened, how often they are used, and other app activity. The linking together of location services data and app interactions is made possible through a unique value that is associated with each mobile device. This value is the **mobile advertising identifier (MAID).** MAID is a unique number that identifies a specific device. Each package of combined user data is identified by a MAID. MAIDs are built into both Google Android and Apple iOS smartphones, as well as a growing number of other devices like game consoles, tablets, and smart TVs.

Note 6

Apple's MAID is called the *Identifier for Advertisers* (*IDFA*) while Google's MAID is known as the *Android Advertising Identifier* (*AAID*).

Who Are the Data Thieves?

Many entities steal data from users. Three of the primary entities are surveillance-based advertisers, governments, and schools.

Surveillance-Based Advertisers

Advertising is communicating with users about a product or service, and *advertisements* are paid messages intended to inform or influence those who receive them. To some degree, all advertising is considered *targeted* to specific groups who would likely be interested in that product or service. This can best be illustrated by mistargeted advertising: an advertisement displayed to individuals living in Tennessee about an automobile dealer in Kansas is mistargeted and is a waste of money for the advertiser and frustration to the ad recipient.

Note 7

Traditional targeted advertising can be broken down by zip code, household income, consumer behaviors, personality traits, lifestyle choices, and general interests. This helps to *slightly* narrow the audience who sees the advertisement to those who may be more likely to purchase the product or service. However, these groups, such as all those who live in a specific zip code, are still very broad.

In traditional marketing, ads are placed in one or a small number of *pre-determined locations*. One example is purchasing ad space in a fitness magazine to reach consumers who purchase athletic accessories (called *contextual ads*). The hope is that interested users read the magazine and see the ad. However, there is no assurance this will take place: an interested user may happen to skip the page with the ad or may not even pick up the magazine itself. Users who are not interested in fitness who see the ad will ignore it because it does not address their interests or needs.

Note 8

The amount of advertising directed to individuals has grown exponentially. In the 1970s, users were exposed to 500 to 1,600 ads per day. By 2007, that number had increased to 5,000 each day. In 2021, a user saw between 6,000 and 10,000 advertisements daily.[1]

Internet advertising has overcome the limitations of traditional advertising. **Surveillance-based advertising** (sometimes also called *ad tech*) refers to Internet-based digital advertising that is targeted at individuals who have been *pre-identified* through smartphone tracking data.

Note 9

Surveillance-based advertising is sometimes called behavioral, personalized, and tailored marketing.

Surveillance-based advertising is different because the ad is targeted at an individual based on the *characteristics* of the individual and not a broad category such as zip code. In many instances, the user has already expressed some type of interest in the product or service. A user searching online for athletic accessories will receive surveillance-based advertising for this product because they have already pre-identified themselves and expressed an interest. Also, these ads are not placed in a pre-determined location with the hope the consumer will see them. Instead, a surveillance-based ad will follow the consumer continually for several days by appearing repeatedly on their smartphone, laptop, and tablet whenever the user opens a web browser.

Note 10

Surveillance-based advertising typically grows 30 to 50 percent annually. It is estimated that two-thirds of total global ad spending is now directed toward these online ads.

Governments

To users living under democratic non-totalitarian governments, it is often surprising to learn of the user data that is routinely accessed by governments without the citizen's knowledge or consent. Data collected by surveillance-based advertisers can essentially be purchased by any entity, including governments. This data can be used to monitor the actions of citizens on an ongoing basis without the citizen's knowledge or permission.

Note 11

Over half the world's population lives under governments that have mandated or are considering mandating government access to data.

Besides purchasing data from surveillance-based advertisers, governments routinely collect their own data. An example is the **Communications Assistance for Law Enforcement Act (CALEA)**, which is a wiretapping law passed in 1994. CALEA's purpose is to enhance the ability of law enforcement agencies to conduct a lawful interception of communications. It requires that telecommunications carriers and manufacturers of telecommunications equipment modify and design their equipment, facilities, and services to ensure that they have built-in capabilities for targeted surveillance, allowing federal agencies to selectively wiretap any telephone traffic. Since 1994, CALEA has since been extended to cover broadband Internet and voice over IP (VoIP) Internet traffic.

Caution 

Some government agencies argue that CALEA covers mass surveillance of communications rather than just tapping specific lines and that not all CALEA-based access requires a warrant.

In addition, "secret" government data collection activities frequently come to light. In mid-2022, a court case revealed that network researchers and security experts inside and outside the U.S. government routinely monitor the world's Internet traffic. Specifically, they perform "data mining" on data related to the Internet's Domain Name System (DNS), which is a kind of "phone book" of the Internet. The court case showed that the government was able to identify a secret communications channel between an American citizen and a foreign nation.

Around this same time, another incident came to light. It was revealed that the Federal Bureau of Investigation (FBI) performed searches of 3.4 million Americans' electronic data in a single year—all without a warrant. (The number of searches the FBI performed the previous year was 1.3 million.) This data was collected not by the FBI but instead by the National Security Agency (NSA) as part of the 1978 Foreign Intelligence Surveillance Act (FISA), which outlines foreign intelligence gathering. FISA allows the NSA to collect intelligence from international phone calls and emails about terrorism suspects, cyber threats, and other security risks. However, much of the data collection also includes data on Americans, such as names, telephone numbers, email addresses, and Social Security numbers.

Note 12

Not all data used by the government is collected by it. One large bank is sharing its customer's saving, spending, and borrowing habits with the government. This includes deposited paychecks, unemployment benefits, and even spending on gasoline purchases. The stated aim is to assist the government in designing its economic policies.

Schools

Educational institutions have increasingly been active in gathering information about their students. Table 6-1 outlines the types of data that schools can collect.

Table 6-1 Data types collected by schools

Data type	Example	Description
Location data	Wi-Fi connections, contactless chips in bus passes, and ID cards	Schools have used this data for automated attendance tracking, class tardiness, and identifying who is riding a school bus.
Audiovisual data	Facial recognition	Locating a student on a large campus can be easily done through monitoring cameras and microphones of recorded images and sounds.
Web browsing data	Visited websites	Schools can monitor which websites a student visits and then intervene if necessary.
Device usage	Social media postings	Offenders who post harmful or threatening messages can be tracked by monitoring their device usage.

The data collected by schools can be used in different ways, such as locating a missing student, determining if a student attended a class or was tardy, or may pose a threat to other students. In some cases, student data is reported to school resource officers or the police.

Caution !

Surveillance-based advertisers, governments, and schools are not the only entities that collect and use data without the user's knowledge and permission. Modern office buildings likewise collect and use employee data. For example, cameras recognize employees' faces when they enter the building and then summon elevators pre-programmed to deliver them only to the specific floor of their office. Wall sensors measure particles and carbon dioxide levels in meeting rooms, monitoring if a virus could be present in the room. Arriving packages are stored in small lockers that employees open with a personal QR code. Some new buildings contain over 300 sensors per floor, continually collecting data on employees.

What Are the Risks?

There are significant risks to the collection and usage of user data. The risks fall into several general categories:

- **Associations with groups**. One common use of personal data is to place what appears to be similar individuals together into groups. One data broker has 70 distinct segments (*clusters*) within 21 consumer and demographic characteristic groups (*life stages*). These groups range from *Boomer Barons* (baby boomer-aged households with high education and income), *Hard Chargers* (well-educated and professionally successful singles), and *True Blues* (working parents who hold blue-collar jobs with teenage children about to heave home). Once a person is placed in a group, the characteristics of that group are applied, such as whether a person is a "potential inheritor," an "adult with senior parent," or whether a household has a "diabetic focus" or "senior needs." However, these assumptions may not always be accurate for the individual that has been placed within that group. Individuals might be offered fewer, or the wrong types of services, based on their association with a group.
- **Statistical inferences**. Statistical inferences are often made that go beyond groupings. For example, researchers have demonstrated that by examining only four data points of credit card purchases (such as the dates and times of purchases) by 1.1 million people, they were able to correctly identify 90 percent of them.[2] In another study, the *Likes* indicated by Facebook users can statistically reveal their sexual orientation, drug use, and political beliefs.[3]
- **Unintended cross-pollination**. "Cross-pollination" is a sharing or interchange of data and is often intended to create a higher benefit than a single data point would. For example, a government-backed initiative to extend credit to people who have traditionally lacked opportunities to borrow money now allows banks to factor in information from the applicant's checking or savings accounts to increase their chances of being approved for credit. This program looks at an applicant's account balances over time and their overdraft histories. Another trend is to allow customers to purchase life-insurance policies without a medical examination by providing insurers access to private medical records. As well-intended as these may be, they can cast a negative light on a person who may have experienced a job loss or a medical issue, and the reasons for the rejection of these services can further impact the individual.
- **Identity theft**. Identity theft involves stealing another person's personal information, such as a Social Security number, and then using the information to impersonate the victim, often for financial gain. The thieves may create new bank or credit card accounts under the victim's name and then charge large purchases to these accounts, leaving the victim responsible for the debts and ruining their credit rating. Almost always, identity theft begins with the theft of personal data that has been collected on a user by surveillance-based advertisers, businesses, governments, schools, or any other entity that gathers user data.

Note 13

Identity theft is covered in Module 1.

- **Individual inconveniences.** Data that has been collected on individuals is most frequently used to direct ad marketing campaigns toward the person. These campaigns include web advertisements, text messages, email, direct mail marketing promotions, and telephone calls. They are often considered annoying and unwanted.

Other issues raised regarding how private data is gathered and used are listed in Table 6-2.

Table 6-2 Issues regarding how private data is gathered and used

Issue	Explanation
The data is gathered and kept secret.	Users have no formal rights to find out what private information is being gathered, who gathers it, or how it is being used.
The accuracy of the data cannot be verified.	Because users do not have the right to correct or control what personal information is gathered, its accuracy may be suspect. In some cases, inaccurate or incomplete data may lead to erroneous decisions made about individuals without any verification.
Identity theft can impact the accuracy of data.	Victims of identity theft will often have information added to their profile that was the result of actions by the identity thieves, and even this vulnerable group has no right to see or correct the information.
Unknown factors can impact overall ratings.	Ratings are often created by combining thousands of individual factors or data streams, including race, religion, age, gender, household income, zip code, presence of medical conditions, transactional purchase information from retailers, and hundreds more data points about individual consumers. How these different factors impact a person's overall rating is unknown.
Informed consent is usually missing or misunderstood.	Statements in a privacy policy such as "We may share your information for marketing purposes with third parties" is not clearly informed consent to freely allow the use of personal data. Often users are not even asked for permission to gather their information.
Data is being used for increasingly important decisions.	Private data is being used on an ever-increasing basis to determine eligibility for significant life opportunities, such as jobs, consumer credit, insurance, and identity verification.
Targeted ads based on private data can lead to discrimination.	Targeted advertising can perpetuate and reinforce harmful stereotypes. For example, research has shown that online employment ads for science, technology, engineering, and mathematics are disproportionately shown to men and hidden from women.

Note 14

User data is often aggregated by *data brokers*, who then sell the data to interested third parties such as marketers or even governments. Unlike consumer reporting agencies, which are required by federal law to give consumers free copies of their credit reports and allow them to correct errors, data brokers are not required to show consumers information that has been collected about them or provide a means of correcting it.

Two Rights & A Wrong

1. Only about 10 percent of smartphone apps have tracking features.
2. As smartphones determine their current location, this information is then packaged into a feature on smartphones known as location services.
3. Surveillance-based advertising is targeted at an individual based on the characteristics of the individual.

See Appendix A for the answer.

Privacy Protections

It is impossible today to completely prevent the collection and use of all private data. Nevertheless, several different protections can reduce the risks associated with other entities that take advantage of users' private data. These protections include using cryptography, limiting cookies, disabling and monitoring MAIDs, and following privacy best practices. There are also responsibilities of organizations towards the collection and usage of user data.

Use Cryptography

Using cryptography involves defining what it is, its benefits, and the types of cryptography. It also involves understanding how cryptography can be used as a tool to protect data.

What Is Cryptography?

As early as 600 BC, the ancient Greeks wrestled with how to keep messages sent by couriers from falling into enemy hands. One early method was to tell the message to the courier so he could later repeat it when he arrived at his destination. However, the disadvantage to this approach was the courier would often paraphrase the message in his own words, which would result in omissions or variations due to his forgetfulness. In some cases, the courier might even intentionally alter the message if he had been bribed or blackmailed by the enemy. Written dispatches, on the other hand, would accurately convey the message and enable it to be sent without the courier knowing its contents, thus reducing the risk of a security breach. But written messages could be seized by the enemy if the courier was captured and searched.

The ancient Greeks turned to making their messages intelligible only to the desired recipient. One technique used was to substitute one letter for another (in English A=M, B=N, C=O, etc.), but this was fairly easy to uncover. The Greeks also marked letters of an ordinary text with tiny dots to indicate which letters, when combined, revealed the secret message. The Greeks even wrote poems in which the first letter of each line of the poetry could be extracted to spell out a message.[4]

By making the message more difficult to read, such as substituting letters, the ancient Greeks were also performing one of the early forms of **cryptography**, which is from Greek words meaning *hidden writing*. Cryptography is the practice of transforming information so that it cannot be understood by unauthorized parties and thus is secure. This is usually accomplished through "scrambling" the information in such a way that only approved recipients (either human or machine) can understand it.

When using cryptography, the process of changing the original text into a scrambled message is known as **encryption** (the reverse process is **decryption** or changing the message back to its original form). In addition, other terminology applies to cryptography:

- *Plaintext.* Unencrypted data that is input for encryption or the output of decryption is called *plaintext.*
- *Ciphertext. Ciphertext* is the scrambled and unreadable output of encryption.
- *Cleartext.* Unencrypted data that is not intended to be encrypted is *cleartext* (it is "in the clear").

Note 15

One of the most famous unsolved cryptographic puzzles in the world is Kryptos. In 1988, the Central Intelligence Agency (CIA) commissioned Maryland-based artist Jim Sanborn to create a sculpture for its expanding headquarters because they wanted to install art that would be relevant to its mission of cracking secrets. The main part of Kryptos is a wavy wall of copper about 20 feet long and 12 feet high (6 meters by 3.6 meters). Into the copper, Sanborn carved a secret message consisting of 1,800 seemingly random letters and four question marks. Thirty years after its dedication in 1990, the Kryptos code has still not been fully cracked, even by the CIA itself. No one knows the solution except Sanborn.

Plaintext data to be encrypted is input into a cryptographic *algorithm* (also called a *cipher*), which consists of procedures based on a mathematical formula. A *key* is a mathematical value entered into the algorithm to produce the ciphertext. Just as a key is inserted into a door lock to lock the door, in cryptography a unique mathematical key is input into the encryption algorithm to "lock" the data by creating the ciphertext. Once the ciphertext needs to be returned to plaintext so the recipient can view it, the reverse process occurs with the decryption algorithm and key to "unlock" it. Cryptographic algorithms are public and well known; however, the individualized key for the algorithm that a user possesses must at all costs be kept secret. The cryptographic process is illustrated in Figure 6-1.

Figure 6-1 Cryptographic process

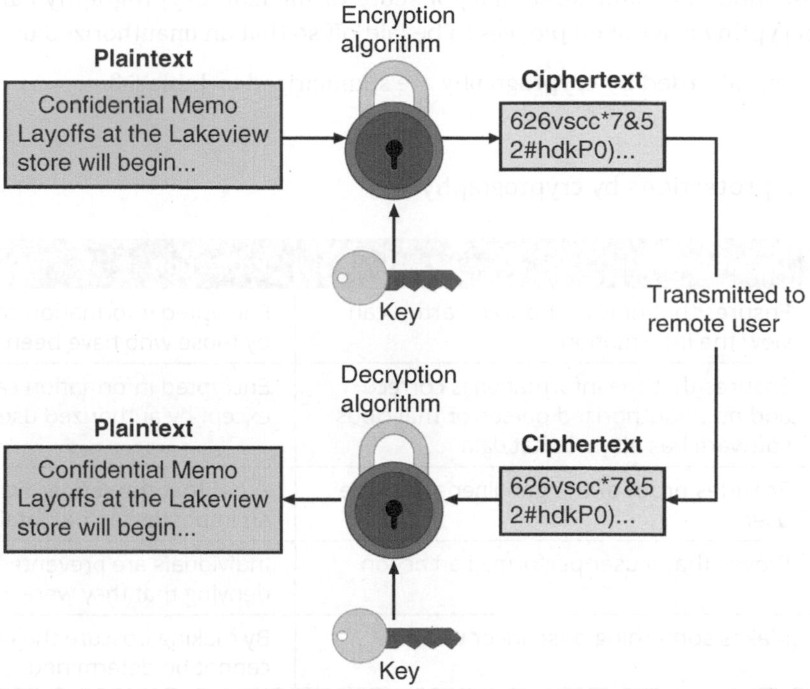

Benefits of Cryptography

Why use cryptography? There are several situations in which cryptography can provide a range of cybersecurity benefits. These protections include:

- *Confidentiality.* Cryptography can protect the confidentiality of information by ensuring that only authorized parties can view it. When private information, such as a list of employees to be laid off, is transmitted across the network or stored on a file server, its contents can be encrypted, which allows only authorized individuals who have the key to read it.
- *Integrity.* Cryptography can protect the integrity of information. Integrity ensures that the information is correct and no unauthorized person or malicious software has altered that data. Because ciphertext requires that a key must be used to open the data before it can be changed, cryptography can ensure its integrity. The list of employees to be laid off, for example, can be protected so that no names can be added or deleted by unauthorized personnel.
- *Authentication.* The authentication of the sender can be verified through cryptography. Specific types of cryptography, for example, can prevent a situation such as the circulation of a list of employees to be laid off that appears to come from a manager, but in reality, was sent by an imposter.
- *Non-repudiation.* Cryptography can enforce non-repudiation. *Repudiation* is defined as denial; non-repudiation is the inability to deny. In information technology, *non-repudiation* is the process of proving that a user performed an action, such as sending an email message. Non-repudiation prevents an individual from fraudulently reneging on an action. The non-repudiation features of cryptography can prevent a manager from claiming he never sent the list of employees to be laid off to an unauthorized third party.

Note 16

A practical example of non-repudiation is Astrid taking her car into a repair shop for service and signing an estimate form of the cost of repairs and authorizing the work. If Astrid later returns and claims she never approved a specific repair, the signed form can be used as non-repudiation.

- *Obfuscation. Obfuscation* is making something obscure or unclear. Cryptography can provide a degree of obfuscation by encrypting a list of employees to be laid off so that an unauthorized user cannot read it.

The security protections afforded by cryptography are summarized in Table 6-3.

Table 6-3 Information protections by cryptography

Characteristic	Description	Protection
Confidentiality	Ensures that only authorized parties can view the information	Encrypted information can only be viewed by those who have been provided the key.
Integrity	Ensures that the information is correct and no unauthorized person or malicious software has altered that data	Encrypted information cannot be changed except by authorized users who have the key.
Authentication	Provides proof of the genuineness of the user	Proof that the sender was legitimate and not an imposter can be obtained.
Non-repudiation	Proves that a user performed an action	Individuals are prevented from fraudulently denying that they were involved in a transaction.
Obfuscation	Makes something obscure or unclear	By making obscure the original information cannot be determined.

Cryptography can protect data as that data resides in any of three states:

- *Data in processing.* **Data in processing** (also called *data in use*) is data on which actions are being performed by devices, such as printing a report from a device.
- *Data in transit.* Actions that transmit the data across a network, like an email sent across the Internet, are called **data in transit** (sometimes called *data in motion*).
- *Data at rest.* **Data at rest** is data that is stored on electronic media.

Types of Cryptography

There are two types of cryptography. The first is symmetric cryptography, while the other is asymmetric cryptography.

Symmetric Cryptography The original cryptography for encrypting and decrypting data is symmetric cryptography. **Symmetric cryptography** uses the same key to encrypt and decrypt the data. Data encrypted by Bob with a key can only be decrypted by Alice using that same key. Thus, it is essential that the key be kept private (confidential), so for this reason, symmetric encryption is also called *private key cryptography*. Symmetric encryption is illustrated in Figure 6-2 where identical keys are used to encrypt and decrypt a document.

Figure 6-2 Symmetric (private key) cryptography

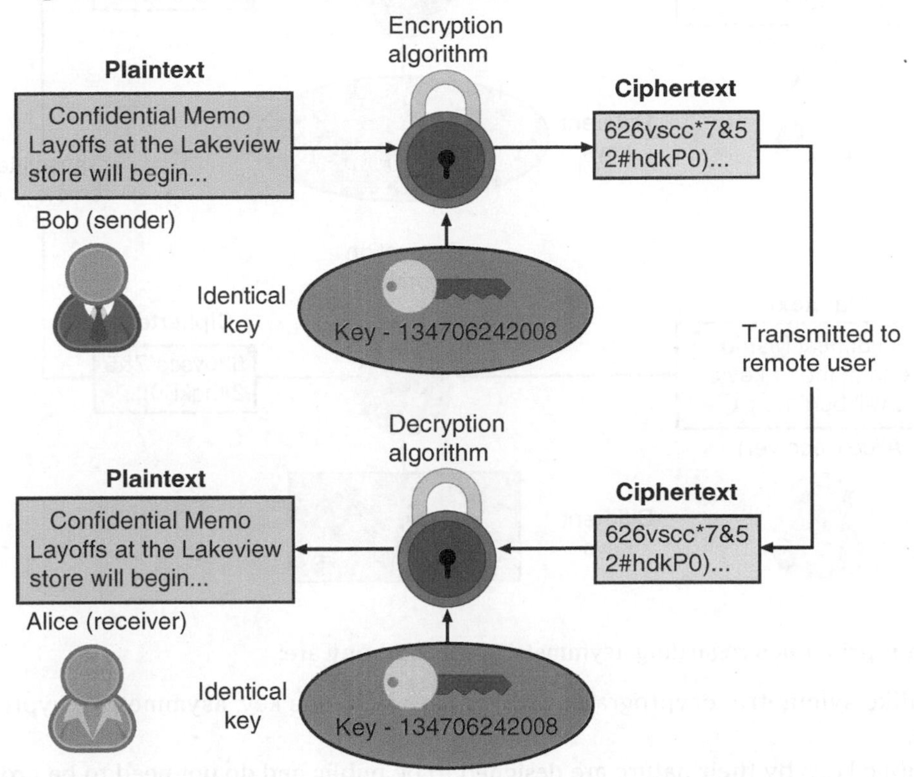

Symmetric cryptography can provide strong encryption—*but only if the key is kept secure between the sender and all the recipients*. If Bob wants to send an encrypted message to Alice using symmetric encryption, he must be sure that she has the key to decrypt the message. Yet how should Bob get the key to Alice? He cannot send it electronically as an email attachment, because that would make it vulnerable to interception by attackers. Nor can he encrypt the key and send it, because Alice would not have a way to decrypt the encrypted key. This illustrates the primary weakness of symmetric cryptography: distributing and maintaining a secure single key among multiple users, who are often scattered geographically, poses significant challenges.

Asymmetric Cryptography A completely different approach is **asymmetric cryptography**, also known as *public key cryptography*. Asymmetric encryption uses *two keys* instead of only one. These keys are mathematically related and are known as the public key and the private key. The **public key** is known to everyone and can be freely distributed, while the **private key** is known only to the individual to whom it belongs. When Bob wants to send a secure message to Alice, he uses Alice's public key to encrypt the message. Alice then uses her private key to decrypt it. Asymmetric cryptography is illustrated in Figure 6-3.

Figure 6-3 Asymmetric (public key) cryptography

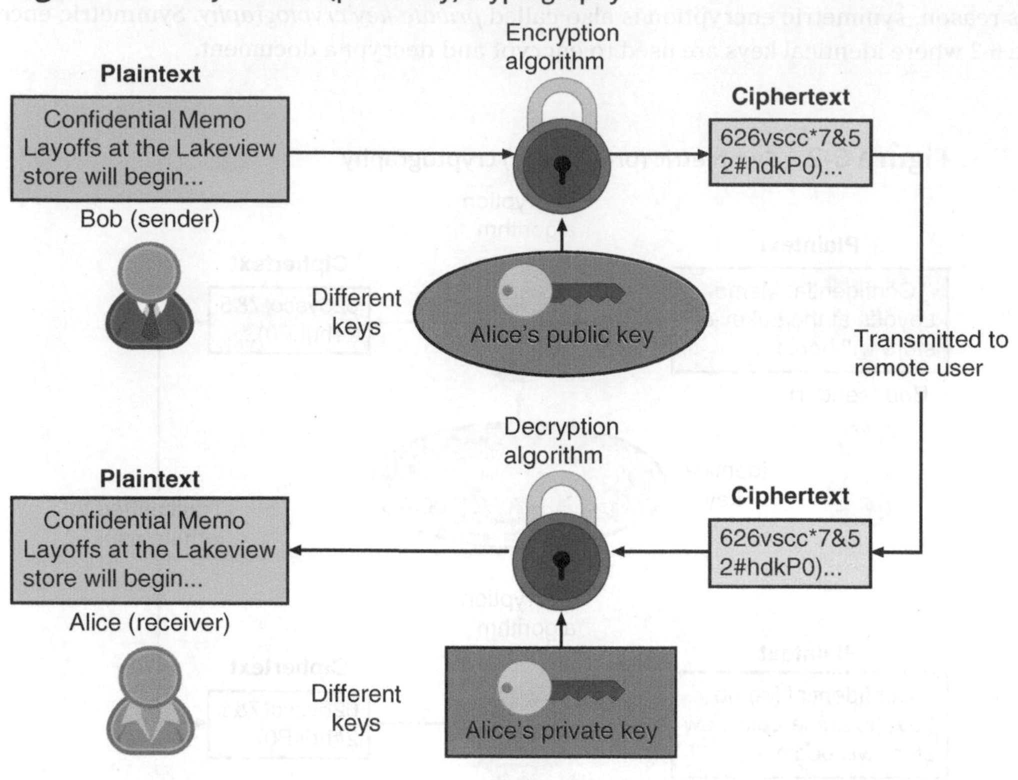

Several important principles regarding asymmetric cryptography are:

- *Key pairs.* Unlike symmetric cryptography, which uses only one key, asymmetric cryptography requires a pair of keys.
- *Public key.* Public keys by their nature are designed to be public and do not need to be protected. They can be freely given to anyone or even posted on the Internet.
- *Private key.* The private key should be kept confidential and never shared.
- *Both directions.* Asymmetric cryptography keys can work in both directions. A document encrypted with a public key can be decrypted with the corresponding private key. In the same way, a document encrypted with a private key can be decrypted with its public key.

Caution !

No user other than the owner should ever have the private key.

Public and private keys may result in confusion regarding whose key to use and which key should be used. Table 6-4 lists the practices to follow when using asymmetric cryptography.

Table 6-4 Asymmetric cryptography practices

Action	Whose key to use	Which key to use	Explanation
Bob wants to send Alice an encrypted message.	Alice's key	Public key	When an encrypted message is to be sent, the recipient's key, and not the sender's key, is used.
Alice wants to read an encrypted message sent by Bob.	Alice's key	Private key	An encrypted message can be read only by using the recipient's private key.

Action	Whose key to use	Which key to use	Explanation
Bob wants to send a copy to himself of the encrypted message that he sent to Alice.	Bob's key	Public key to encrypt Private key to decrypt	An encrypted message can be read only by the recipient's private key. Bob would need to encrypt it with his public key and then use his private key to decrypt it.
Bob receives an encrypted reply message from Alice.	Bob's key	Private key	The recipient's private key is used to decrypt received messages.
Bob wants Susan to read Alice's reply message that he received.	Susan's key	Public key	The message should be encrypted with Susan's key for her to decrypt and read with her private key.

Protections Through Cryptography

Cryptography can secure private data by making it inaccessible to anyone who does not have the key. Protections through cryptography can be categorized as it pertains to protecting data at rest and data in transit.

Data at Rest Data at rest is data that is stored on electronic media on a mobile device, laptop, or external storage unit. Cryptography can be implemented for data at rest through software running on a device. Encryption can also be performed on a larger scale by encrypting the entire storage unit itself.

Encryption Through Software Cryptographic software can be used to encrypt or decrypt files one by one. However, this can be a cumbersome process. Instead, protecting groups of files, such as all files in a specific folder, can take advantage of the operating system's (OS's) file system (a *file system* is a method used by OSs to store, retrieve, and organize files). There are a wide variety of third-party software tools available for performing encryption. These include GNU Privacy Guard (which is abbreviated GNuPG), AxCrypt, Folder Lock, and VeraCrypt, which is seen in Figure 6-4.

Figure 6-4 VeraCrypt

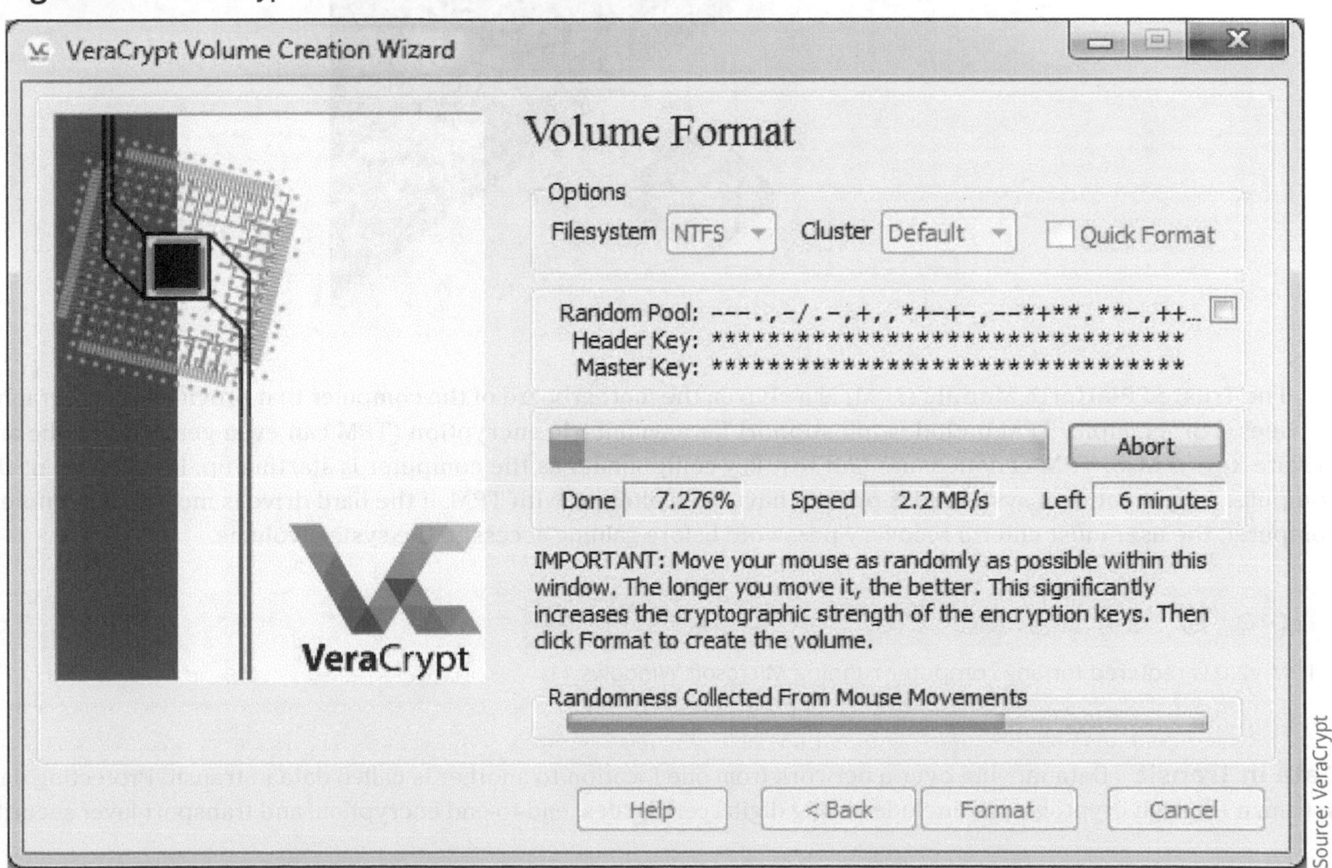

Source: VeraCrypt

Modern OSs provide encryption support natively. Microsoft's *Encrypting File System* (*EFS*) is a cryptography system for Windows, while Apple's *FileVault* performs a similar function. Because these are tightly integrated with the file system, file encryption and decryption are transparent to the user. Any file created in an encrypted folder or added to an encrypted folder is automatically encrypted. When an authorized user opens a file, it is decrypted as data is read from a disk; when a file is saved, the OS encrypts the data as it is written to a disk.

Cryptography can also be applied to entire storage units instead of individual files or groups of files. This is known as **full disk encryption (FDE)** and protects all data on a storage unit. One example of full disk encryption software is that included in Microsoft Windows known as *BitLocker* drive encryption software. BitLocker prevents attackers from accessing data by booting from another OS or placing the storage unit in another computer.

Hardware Encryption Software encryption suffers from the same fate as any application program: it can be subject to attacks to exploit its vulnerabilities. As a more secure option, cryptography can be embedded in hardware that cannot be exploited like software encryption.

Self-encrypting drives (SEDs) can protect all files stored on them. When the computer or other device with an SED is initially powered up, the drive and the host device perform an authentication process. If the authentication process fails, the drive can be configured to simply deny any access to the drive or even perform a *cryptographic erase* on specified blocks of data (a cryptographic erase deletes the decryption keys so that no data can be recovered). This also makes it impossible to install the drive on another computer to read its contents.

A **Hardware Security Module (HSM)** is a removable external cryptographic hardware device. HSMs are popular consumer-level devices. Figure 6-5 shows a USB consumer HSM. Some financial banking software comes with a specialized HSM hardware key, also called a "security dongle." This device is paired with a specific financial account and cannot be cloned or compromised.

Figure 6-5 USB HSM

Source: Yubica

The **Trusted Platform Module (TPM)** is a chip on the motherboard of the computer that provides cryptographic services. For example, TPM includes full support for asymmetric encryption (TPM can even generate public and private keys). Also, TPM can measure and test key components as the computer is starting up. It will prevent the computer from booting if system files or data have been altered. With TPM, if the hard drive is moved to a different computer, the user must enter a recovery password before gaining access to the system volume.

Note 17

TPM v2.0 is required for any computer running Microsoft Windows 11.

Data in Transit Data moving over a network from one location to another is called data in transit. Protecting data in transit through cryptography includes using digital certificates, end-to-end encryption, and transport layer security.

Digital Certificates Suppose that Alice is making a purchase at an online e-commerce site and needs to enter her credit card number. How can she be certain that she is at the authentic website of the retailer and not an imposter's lookalike site that will steal her credit card number?

The solution is a digital certificate. A **digital certificate** is a technology used to associate a user's identity to a public key and that has been digitally signed by a trusted third party. This third party verifies the owner and that the public key belongs to that owner. A digital certificate is basically a container for a public key. Typically, a digital certificate contains information such as the owner's name or alias, the owner's public key, the name of the issuer, the digital signature of the issuer, the serial number of the digital certificate, and the expiration date of the public key. It can contain other user-supplied information, such as an email address, postal address, and basic registration information.

Note 18

When Bob sends a message to Alice, he does not ask her to retrieve his public key from a central site. Instead, Bob attaches the digital certificate to the message. When Alice receives the message with the digital certificate, she can check the signature of the trusted third party on the certificate. If the signature was signed by a party that she trusts, then Alice can safely assume that the public key—contained in the digital certificate—is actually from Bob. Digital certificates make it possible for Alice to verify Bob's claim that the key belongs to him and prevent an attack that impersonates the owner of the public key.

The online retailer's web server issues to Alice's web browser a digital certificate that has been signed by a trusted third party. In this way, Alice can rest assured that she is at the authentic online retailer's site.

There are several different types of digital certificates. Two of the more common include:

- **Code signing digital certificate**. Digital certificates are used by software developers to digitally sign a program to prove that the software comes from the entity that signed it and no unauthorized third party has altered or compromised it. This is known as a **code signing digital certificate**. When the installation program is launched that contains a code-signed digital certificate, a pop-up window appears that says *Verified publisher* as seen in Figure 6-6. An installation program that lacks a code digital certificate will display a window with the warning *Publisher:Unknown*.

Figure 6-6 Verified publisher message

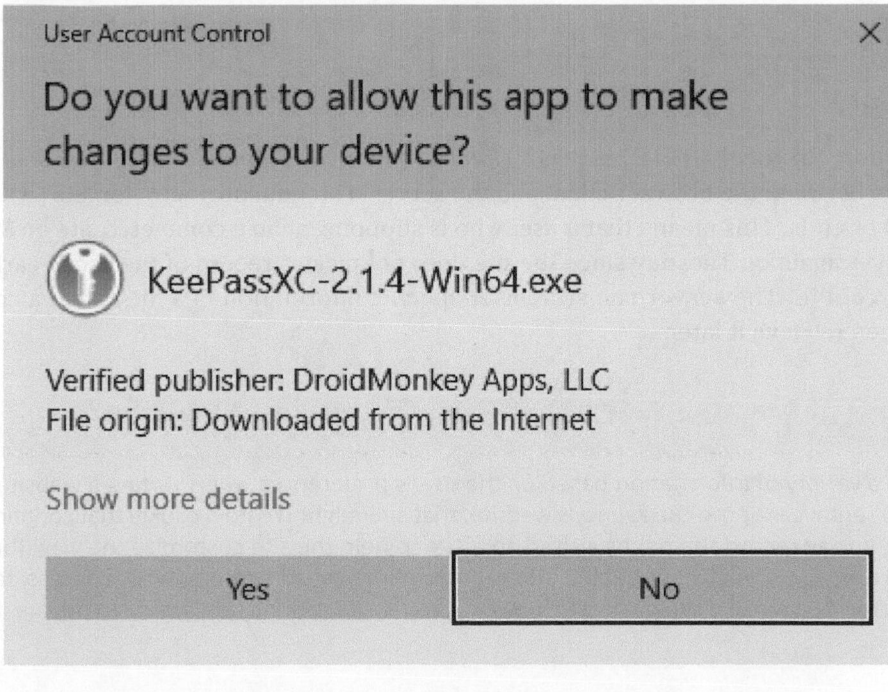

Source: KeePassXC Team

- **Email digital certificate**. An **email digital certificate** allows a user to digitally sign and encrypt mail messages. Typically, only the user's name and email address are required to receive this certificate.

End-to-End Encryption **End-to-end encryption (E2EE)** protects messages in transit all the way from the sender to the receiver. This ensures that information to which cryptography is applied (encrypted) by the sender (the first "end") can *only* be decrypted by its final recipient (the second "end"). No one, including the company behind the app that is being used, can "listen in" and eavesdrop on the communication.

Transport Layer Security **Transport Layer Security (TLS)** protects messages *only* as they travel from the user's device to the app's servers, and then from the app's servers to the recipient's device. In the middle, the messaging service provider or the website that is being visited *can* view unencrypted copies of the messages. Because these messages can be seen by and are often stored on company servers, they may be vulnerable to law enforcement requests or leakage if the company's servers are compromised.

One example of TLS is the Hypertext Transport Protocol Secure (HTTPS). Suppose that Alice is making a purchase at an online e-commerce site and needs to enter her credit card number. That number is protected while in transit *to* the e-commerce web server, but once it arrives, the card number must be able to be read and then processed by the web server.

Note 19

HTTPS is covered in Module 4.

Another example of TLS is a virtual private network (VPN). A VPN sends encrypted traffic between the sender and the VPN provider. If someone were to spy on the local network to attempt to see the websites that are being visited, they will be able to see that that user is connected to a VPN but not be able to see what websites are being viewed. However, the VPN provider *can* view all user traffic. In many instances, the VPN provider will store that information and make it available to law enforcement agencies or it could be susceptible to a data breach.

Note 20

VPNs are covered in Module 5.

Limit Cookies

The Hypertext Transport Protocol (HTTP) is it is a *stateless protocol*. Unlike a *stateful protocol* that "remembers" everything that occurs between the browser client and the server, a stateful protocol "forgets" what occurs when the session is interrupted or ends. This means that a user who is shopping at an e-commerce site on Monday using HTTP will have to start all over again on Tuesday since the site does not make a record of the user's earlier shopping.

One solution is a **cookie**. The server can store user-specific information in a file called a cookie on the user's local computer and then retrieve it later.

Note 21

A cookie can contain a variety of information based on the user's preferences when visiting a website. For example, if a user inquires about a rental car at the car agency's website, that site might create a cookie that contains the user's travel itinerary. In addition, it may record the pages visited on a site to help the site customize the view for any future visits. Cookies can also store any personally identifiable information (name, email address, work address, telephone number, and so on) that was provided when visiting the site; however, a website cannot gain access to private information stored on the local computer.

There are several types of cookies. A **first-party cookie** is created from the website that a user is currently viewing; whenever the user returns to this site, that cookie would be used by the site to view the user's preferences and better customize the browsing experience. Some websites attempt to place additional cookies on the local hard drive. These cookies often come from third parties that advertise on the site and want to record the user's preferences and are called **third-party cookies**. A **session cookie** is stored in random access memory (RAM), instead of on the hard drive, and only lasts for the duration of visiting the website.

Cookies can pose security as well as privacy risks. First-party cookies can be stolen and used to impersonate the user, while third-party cookies can be used to track the browsing or buying habits of a user. When multiple websites are serviced by a single marketing organization, cookies can be used to track browsing habits on all the client's sites. Clearing the browser's memory (*cache*) of cookies can help ensure more privacy, but it can impact the browsing and online shopping experience.

Note 22

As a means of protection for cookies, a web browser can send a *secure cookie*. This cookie is only sent to the server with an encrypted request over the secure HTTPS protocol. This prevents an unauthorized person from intercepting a cookie that is being transmitted between the browser and the web server.

Disable and Monitor MAIDs

For mobile devices, MAIDs can be disabled to prevent tracking. For Google Android 12 devices, open the **Settings** app and navigate to **Privacy > Ads**. Tap **Delete advertising ID** and then tap it again on the next page to confirm the selection. This will prevent any app on the device from accessing it in the future.

Apple now requires apps to ask permission before an app can track using the MAID. When installing a new app, users may be asked for permission to track. Selecting **Ask App Not to Track** will deny access to the MAID. To monitor which apps already have access, go to **Settings > Privacy > Tracking**. Tracking for individual apps that have previously received permission can be disabled. In addition, Apple has created its own targeted advertising system separate from third-party apps. To disable the Apple system, navigate to **Settings > Privacy > Apple Advertising**. Set **Personalized Ads** to the **off** position.

Caution !

These steps may vary as updates to Android and iOS are released.

Follow Privacy Best Practices

To protect important information, users should consider the following privacy best practices:

- Use encryption to protect sensitive documents that contain personal information, such as a Social Security number, driver's license number, bank account numbers, etc. Store the encryption keys in a password manager.
- Be sure that strong passwords are used on all accounts that contain personal information.
- Shred financial documents and other paperwork that contains personal information before discarding it.

Note 23

There are three common types of shredders. A strip-cut device shreds paper into long vertical strips. A cross-cut shredder adds horizontal cuts to make the shredded pieces even smaller while a micro-cut device shreds documents into tiny pieces. Cross-cut and micro-cut shredders offer better security than a strip-cut shredder.

- Do not carry a Social Security number in a wallet or write it on a check.
- Do not provide personal information either over the phone or through an email message.
- Keep personal information in a secure location in a home or apartment.
- Be cautious about what information is posted on social networking sites and who can view your information. Show "limited friends" a reduced version of a profile, such as casual acquaintances or business associates.
- Keep only the last three months of the most recent financial statements and then shred older documents instead of tossing them in the trash or a recycling bin. For paper documents that must be retained, use a scanner to create a PDF of the document and then add a strong password to the PDF file that must be entered before it can be read.
- Use the private browsing option available in most browsers to limit the collection of data. When not using private browsing, delete the browsing history and clear the cache after each session.
- Review the privacy options of the web browser and turn on those features that will provide the highest level of privacy without negatively impacting the browser experience.
- Turn on Wi-Fi Protected Access (WPA2 or WPA3) Personal on Wi-Fi networks to prevent an unauthorized person from viewing wireless transmissions.

Note 24

WPA2 and WPA3 Personal are covered in Module 5.

- Give cautious consideration before permitting a website or app request to collect data.
- Be sure that *https* appears at the beginning of a web address that asks for credit card numbers or other sensitive information. Do not provide any information if that is not present.
- Be cautious about surrendering personal information in exchange for a coupon or to enter a contest.

Caution !

Due to restrictions on MAIDs by Apple and Google, marketers of consumer products have taken action to bypass MAID and collect their own information on consumers. Today, brands are deploying an array of tactics to persuade users to surrender data to the brand itself through loyalty programs, sweepstakes, newsletters, quizzes, polls, and QR codes. One marketing organization that represents avocado growers and packers is encouraging customers to submit grocery receipts to earn points exchangeable for avocado-themed sportswear and is also conducting a contest to win a truck. To enter, consumers scan QR codes on in-store displays and enter their name, birthday, email, and phone number.

- Use common sense. Websites that request more personal information than would normally be expected, such as a user name and password to another account, should be avoided.
- Encourage vendors to change their practices on data collection.

Note 25

Pressure from external sources has largely been responsible for vendors like Google and Apple changing their practices. After a 2010 newspaper investigation of Apple followed by questions raised by U.S. members of Congress, Apple began restricting access to its MAID. Google soon began to change its practices as well.

- Advocate for state and federal regulations that limit the collection and usage of private data.

Note 26

Despite the best intentions and diligent actions of users to protect the privacy of their data, sometimes nefarious actions by businesses can counteract these. Twitter agreed to pay a $150 million penalty for targeting ads at users by using their telephone numbers and email addresses that were collected from those users when they enabled two-factor authentication (2FA). The smartphone app for the Tim Hortons restaurant chain asked users for permission for the app to access the mobile device's location services but misled users to believe that it would only be accessed when the app was in use. However, the app tracked users' geolocation whenever the smartphone was turned on.

Responsibilities of Organizations

Organizations that collect users' personal data likewise have responsibilities and obligations. These are summarized in Table 6-5 with actual examples of misuse by organizations, by the responsible action the organization should have taken, and an explanation of the practices.

Table 6-5 Privacy responsibilities of organizations

Example of misuse	Responsible action	Explanation
During the online registration process, the organization required new users to provide both their email address and the password to that email account and then stored the information in cleartext.	Collect only necessary personal information.	Organizations should not collect any personal information unless it is necessary, and the information that is collected should be limited.
An organization collected customers' credit and debit card information to process transactions in its retail stores but then stored that information for 30 days, long after the sale was complete.	Keep personal information only as long as necessary.	Unless there is a legitimate business need, personal information should be securely disposed of as soon as any transactions are completed.
An organization used actual personal information in employee training sessions and then failed to remove the information from employees' computers after the training was completed.	Do not use personal information when it is not necessary.	Fictitious information should be used for any training or development purposes.
Over 7,000 files containing users' personal information were inadvertently sent to a third party by an organization that had failed to restrict employee access to sensitive personal information.	Restrict access to sensitive information.	If employees do not need to use customers' personal information as part of their job function, access to such information should be denied.
An organization gave all of its employees' administrative control over the system, including the ability to reset user account passwords and view users' comments.	Limit administrative access.	Administrative access, which allows a user to make system-wide changes, should be limited to employees who have that job function.
An organization stored sensitive customer information that was encrypted with a nonstandard and proprietary form of encryption, which contained several vulnerabilities.	Use industry-tested and accepted methods.	Organizations should take advantage of the "collected wisdom" of encryption algorithms that have been tested by experts over many years.
Sensitive personal information was thrown away in dumpsters and hard drives that contained personal information were sold as surplus.	Dispose of sensitive data securely.	When paperwork or equipment containing personal information is no longer needed, it should be destroyed by shredding, burning, or pulverizing to make the data unreadable.

Two Rights & A Wrong

1. A key is a mathematical value entered into the algorithm to produce the ciphertext.
2. Symmetric cryptography uses one key to encrypt data and a different key to decrypt data.
3. TLS protects messages only as they travel from the user's device to the app's servers, and then from the app's servers to the recipient's device.

See Appendix A for the answer.

Module Summary

- Every time a user interacts with technology, they leave behind a "data trail," which is a digital record of their activity. As more activities of a user's everyday life are performed electronically using more interconnected devices, the volume of data compiled on each user grows exponentially each day. Today the overwhelming majority of data that is exfiltrated through trackers is from users' smartphones. Smartphone data exfiltration is primarily based on two smartphone features that are part of all smartphones today: location services and mobile advertising identifiers (MAIDs). As smartphones determine their current location, this information is then packaged into a feature on smartphones known as location services. Apps on the smartphone can request the current location data and then can use it to provide services that are based on location. The linking together of location services data and app interactions is made possible through a unique value that is associated with each mobile device known as the mobile advertising identifier (MAID).
- In traditional marketing, ads are placed in one or a small number of pre-determined locations. Surveillance-based advertising (sometimes also called ad tech) refers to Internet-based digital advertising that is targeted at individuals who have been pre-identified through smartphone tracking data. Surveillance-based advertising is different because the ad is targeted at an individual based on the characteristics of the individual. These ads are not placed in a pre-determined location with the hope the consumer will see them but will follow the consumer continually for several days by appearing repeatedly on their smartphone, laptop, and tablet whenever the user opens a web browser.
- Data collected by surveillance-based advertisers can essentially be purchased by any entity, including governments. This data can be used to monitor the actions of citizens on an ongoing basis without the citizen's knowledge or permission. Besides purchasing data from surveillance-based advertisers, governments routinely collect their own data. In addition, "secret" government data collection activities frequently come to light. Educational institutions have increasingly been active in gathering information about their students. The data collected by schools can be used in different ways, such as locating a missing student, determining if a student attended a class or was tardy, or may pose a threat to other students. In some cases, student data is reported to school resource officers or the police.
- There are significant risks to the collection and usage of user data. The risks fall into several general categories such as associations with groups, statistical inferences, unintended cross-pollination, identity theft, and individual inconveniences.
- While it is impossible today to completely prevent the collection and use of all private data, several protections can reduce the risks associated with other entities that take advantage of users' private data. One protection is to use cryptography. Cryptography is the practice of transforming information so that it cannot be understood by unauthorized parties and thus is secure. When using cryptography, the process of changing the original text into a scrambled message is known as encryption. The reverse process is decryption or changing the message back to its original form. The benefits of cryptography include confidentiality, integrity, authentication, non-repudiation, and obfuscation. Cryptography can protect data as that data resides in any of three states: data in processing, data in transit, and data at rest.

- Symmetric cryptography uses the same key to encrypt and decrypt the data. Because it is essential that the key be kept private (confidential), symmetric encryption is also called private key cryptography. The primary weakness of symmetric cryptography is distributing and maintaining a secure single key among multiple users, who are often scattered geographically, which poses significant challenges. A completely different approach is asymmetric cryptography, also known as public key cryptography. Asymmetric encryption uses two keys instead of only one. These keys are mathematically related and are known as the public key and the private key.
- Cryptography can be implemented for data at rest through software running on a device. Encryption can also be performed on a larger scale by encrypting the entire disk drive itself. Protecting data in transit through cryptography includes using digital certificates. A digital certificate is a technology used to associate a user's identity to a public key and that has been digitally signed by a trusted third party. There are several different types of digital certificates.
- End-to-end encryption (E2EE) protects messages in transit all the way from the sender to the receiver. Transport Layer Security (TLS) protects messages only as they travel from the user's device to the app's servers, and then from the app's servers to the recipient's device. Another protection is to limit web browser cookies. A server can store user-specific information in a file on the user's local computer and then retrieve it later in a file called a cookie.
- For mobile devices, MAIDs can be disabled to prevent tracking. To protect important information, users should consider the following privacy best practices. Organizations that collect users' personal data likewise have responsibilities and obligations.

Key Terms

asymmetric cryptography
code signing digital certificate
Communications Assistance for Law Enforcement Act (CALEA)
cookie
cryptography
data at rest
data in processing
data in transit
decryption
digital certificate
email digital certificate
encryption
end-to-end encryption (E2EE)
first-party cookie
full disk encryption (FDE)
Hardware Security Module (HSM)
location services
mobile advertising identifier (MAID)
private key
public key
self-encrypting drives (SEDs)
session cookie
surveillance-based advertising
symmetric cryptography
third-party cookie
Transport Layer Security (TLS)
Trusted Platform Module (TPM)

Review Questions

1. Which of the following is NOT a risk associated with the use of private data?
   a. Individual inconveniences
   b. Statistical inferences
   c. Devices being infected with malware
   d. Associations with groups

2. What is ciphertext?
   a. A mathematical value entered into an algorithm
   b. Data that has been encrypted
   c. The public key of a symmetric cryptographic process
   d. Procedures based on a mathematical formula used to encrypt and decrypt data

3. What is data called that is to be encrypted by inputting it into a cryptographic algorithm?
   a. Opentext
   b. Cleartext
   c. Plaintext
   d. Transparenttext

4. Which of these is NOT a basic security protection for information that cryptography can provide?
   a. Authenticity
   b. Integrity
   c. Risk loss
   d. Confidentiality

5. Rowan needs proof that one of his team members sent an email message to him. What does Rowan need?
- **a.** Repudiation
- **b.** Nonrepudiation
- **c.** Availability
- **d.** Integrity

6. How many keys are used in asymmetric cryptography?
- **a.** One
- **b.** Two
- **c.** Three
- **d.** Four or more

7. Which of these is NOT a method for encryption through hardware?
- **a.** FDE
- **b.** SED
- **c.** HSM
- **d.** TPM

8. If Bob wants to send a secure message to Alice using an asymmetric cryptographic algorithm, which key does he use to encrypt the message?
- **a.** Alice's private key
- **b.** Alice's public key
- **c.** Bob's public key
- **d.** Bob's private key

9. What is the most important advantage of hardware encryption over software encryption?
- **a.** Software encryption cannot be used on older computers.
- **b.** Hardware encryption is 20 times faster than software encryption.
- **c.** Software that performs encryption can be subject to attacks.
- **d.** There are no advantages of hardware encryption over software encryption.

10. Which of these appears in the web browser when you are connected to a secure website that is using a digital certificate?
- **a.** HTTPX://
- **b.** Fingerprint
- **c.** Padlock
- **d.** Red triangle

11. Which of the following is NOT a privacy best practice?
- **a.** Use the private browsing option in your web browser.
- **b.** Shred financial documents and paperwork that contains personal information before discarding it.
- **c.** Carry your Social Security card with you so that it cannot be stolen when you are not home.
- **d.** Use strong passwords on all accounts that contain personal information.

12. Which of these is NOT a responsibility of an organization regarding user private data?
- **a.** Collect only necessary personal information.
- **b.** Use industry-tested and accepted methods.
- **c.** Keep personal information for no longer than 365 days.
- **d.** Do not collect and use personal information when it is not necessary.

13. How many trackers does the average app contain?
- **a.** One
- **b.** Two
- **c.** None
- **d.** Six

14. When entering a password or email address in a web form, when is that data collected?
- **a.** As the data is being entered.
- **b.** When the *Submit* button is clicked.
- **c.** When the webpage is closed.
- **d.** When the user logs out of their computer.

15. Which of the following is NOT true about smartphones?
- **a.** Smartphones are designed to protect users' privacy.
- **b.** It is almost impossible to install a different operating system.
- **c.** Smartphones have become the primary communications tool for accessing the Internet.
- **d.** It is hard to remove or replace potentially unwanted programs (PUPs) that are installed by default.

16. Which of the following is NOT true about the mobile advertising identifier (MAID)?
- **a.** MAID is a unique number that identifies a specific device.
- **b.** Each package of combined user data is identified by a MAID.
- **c.** MAIDs are built into Google Android but not Apple iOS smartphones.
- **d.** Devices like game consoles, tablets, and smart TVs often have MAIDs.

17. Which of the following is NOT true about surveillance-based advertising?
- **a.** Surveillance-based advertising is different because the ad is targeted at an individual based on the characteristics of the individual and not a broad category such as zip code.

b. Surveillance-based advertising ads are not placed in a pre-determined location.
c. By law, surveillance-based advertising can only appear to a user for a maximum of seven days.
d. Most advertising today is surveillance-based advertising.

18. What does the Communications Assistance for Law Enforcement Act (CALEA) address?
a. Wiretapping
b. Surveillance-based advertising
c. MAIDs
d. VXRs

19. Which of the following is NOT a data type collected by schools?
a. Location data
b. Web browsing data
c. Email correspondence
d. Device usage

20. Which of the following is NOT an issue regarding how private data is gathered and used?
a. The data is gathered and kept secret.
b. The accuracy of the data cannot be verified.
c. Informed consent is usually missing or misunderstood.
d. Data is no longer used for important decisions.

Hands-On Projects

Project 6-1: Web Browser Privacy Settings

In this project, you will view web browser privacy settings using a Google Chrome web browser.

1. Open a Chrome web browser and go to **www.cengage.com**.
2. Note that although you did not enter *https://*, nevertheless Google created a secure connection. Why would it do that? What are the advantages?
3. Click the padlock icon in the browser address bar to open the window about this connection.
4. Under **Cookies**, how many cookies are used with this site? Does this number surprise you?
5. Click **Cookies** and if necessary, click the **Allowed** tab.
6. Scroll down the list of cookies that were set from visiting this page. Are any of the names familiar to you?
7. Click the **Blocked** tab. Are any cookies being blocked by this browser?
8. Click **Done**.
9. Click the padlock icon in the browser address bar again to open the window about this connection.
10. Note that the information provided says **Connection is secure**.
11. Click **Connection is secure** and read what this means. How should you use this when visiting a website that asks for information such as a credit card number?
12. Click **Certificate is valid** to view the digital certificate information for this site.
13. Read the **Certificate Information** about this digital certificate.
14. Click the **Details** tab and look at each of the fields and their values.
15. Click **Public key** to display the public key associated with this website. How is the public key used?
16. Click **Certification Path**. What third-party entity issued this digital certificate? What is the status of this certificate?
17. Click **OK**.
18. Click the three dots on the right side of the web browser and then click **Settings**.
19. Click **Privacy and security** in the left pane.
20. Click **Get started** under the **Take the Privacy Guide.**
21. Read through all of the screens. Is this information helpful? Click **Done** when finished.
22. Click **Check now** under **Safety check**. Note the different available safety settings.
23. Click **Cookies and other site data**. Note that different cookie settings can be selected here. Which settings would you choose? Why?
24. Click **See all cookies and site data**. Scroll through the list of cookies that are on this computer. Does the number surprise you?
25. What did you learn from examining these web browser settings? What changes would you make for improved privacy?
26. Close all windows.

Continued

Project 6-2: Using a Non-Persistent Web Browser

Non-persistence tools are those that are used to ensure that unwanted data is not carried forward but instead a clean image is used. This helps protect user privacy. One common tool is a web browser that retains no information such as cookies, history, passwords, or any other data and requires no installation but runs from a USB flash drive. In this project, you download and install a non-persistent web browser.

1. Use your web browser to go to **www.browzar.com** (if you are no longer able to access the program through this URL, use a search engine and search for "Browzar").
2. Click **Key Features** and read about the features of Browzar.
3. Click **Help & FAQs** and read the questions and answers.
4. Click **Download now – it's FREE!**
5. Choose one of the available themes and click **Download**.
6. Click **Accept**.
7. Click **Download**.
8. Click the downloaded file to run Browzar. Note that no installation is required and it can be run from a USB flash drive.
9. From Browzar, go to **www.google.com**.
10. Enter **Cengage** in the search bar to search for information about Cengage.
11. Now click the red **X** in the upper-right corner to close the browser. What information appears in the pop-up window? What happened when you closed the browser?
12. Launch Browzar again.
13. Click **Tools**.
14. Click **Secure delete**.
15. Click **More >>**. What additional protections does Secure Delete give?
16. Close all windows.

Project 6-3: Using 7-Zip Cryptography

There are a wide variety of third-party software tools available for performing encryption and decryption. In this project, you will download and use 7-Zip.

1. Use your web browser to go to **www.7-zip.org/index.html** (if you are no longer able to access the site through the web address, use a search engine to search for "7-zip").
2. Click the appropriate version and click **Download**.
3. Follow the instructions to install the program.
4. Locate a file or a folder on the computer to create an encrypted archive file.
5. Single-click the file and then right-click it. Note that the shortcut menu now has a **7-Zip** option.
6. Click **7-Zip**.
7. Click **Add to Archive**.
8. The **Add to Archive** window appears. Be sure the **Archive format:** is **7z**.
9. Under **Encryption**, enter a strong password.
10. Enter the password again under **Reenter password**.
11. Click **OK**.
12. A new encrypted file is created with the extension *7z*.
13. Now open this encrypted file. Select the file by clicking it.
14. Right-click the selected file and click **7-Zip**.
15. Click **Extract Files**.
16. Under **Password**, enter the password and then click **OK**. The encrypted file will be extracted and available for use.
17. Close all windows.

Project 6-4: Using Microsoft's Encrypting File System (EFS)

Microsoft's Encrypting File System (EFS) is a cryptography system for Windows operating systems. Because EFS is tightly integrated with the file system, file encryption and decryption are transparent to the user. In this project, you will turn on and use EFS.

1. Create a Word document with the contents of the first two paragraphs under **Cybersecurity Headlines** on the first page of this module.
2. Save the document as **Encrypted.docx**.
3. Save the document again as **Not Encrypted.docx**.
4. Close Word.
5. Right-click the **Start** button and then click **File Explorer**.
6. Navigate to the location of **Encrypted.docx**.
7. Right-click **Encrypted.docx**.
8. Click **Properties**.
9. Click the **Advanced** button.
10. Check the box **Encrypt contents to secure data**. This document is now protected with EFS. All actions regarding encrypting and decrypting the file are transparent to the user and should not noticeably affect any computer operations. Click **OK**.
11. Click **OK** to close the Encrypted Properties dialog box.
12. Launch Microsoft Word and then open **Encrypted.docx**. Was there any delay in the operation?
13. Now open **Not Encrypted.docx**. Was it any faster or slower?
14. Close all windows.

Case Projects

Case Project 6-1: Survey About Privacy

Create a short survey to administer to at least three of your friends about what they know regarding privacy as it relates to their smartphones. Include questions such as, "Do you know what location services are?" "What is a MAID?" "Are you aware of surveillance-based advertisers?" "Does your school monitor your actions and how do they use that information?" Write an analysis of their answers and perceptions about privacy.

Case Project 6-2: Current State of Privacy

Regulators and government entities continue to pressure Apple and Google to make changes to their smartphones and user data provided by surveillance-based advertisers. Use the Internet to research current laws and regulations that address privacy. Are these adequate? Should more be done to protect users? What would you recommend? Write a one-page paper on your research.

Case Project 6-3: History of Apple Privacy Protections

Research how Apple has changed how it protects user privacy on smartphones since 2010. Why have they made changes? Are these sufficient? Do Apple iPhones provide better privacy than Google Android devices? Why or why not? Write a one-page paper on your research.

Continued

Case Project 6-4: VPN Risks

Because VPNs use TLS instead of end-to-end encryption, the vendors behind the VPNs can store and then make available the list of websites that users access, which is the very opposite of the goals of a VPN. Research the history of VPNs that have compromised user security by storing this information and then making it available. Next, select three VPNs and carefully read the Terms of Service document. Do they specifically state that no user data will be stored on their servers? How could this be verified? Based on your research, which VPN provider would you choose? Why? Write a one-page paper on your research.

Case Project 6-5: Privacy Best Practices

Review each of the suggested privacy best practices outlined in this module. Give each suggestion two scores, one based on the strength of the privacy (on a scale from 1–5) that it would provide and the other score (on a scale from A–F) of how likely you would be to follow this suggestion and why or why not. Then ask three of your friends to likewise score these suggestions. Finally, design at least three additional privacy best practices of your own. Write a one-page paper on what you have learned.

References

1. Wess, Sydney, "What is targeted advertising?" *The Manifest*, Jan. 12, 2022, accessed May 21, 2022, https://themanifest.com/ppc/blog/what-is-targeted-advertising.
2. Hardesty, Larry, "Privacy challenges," *MIT News*, Jan. 29, 2015, accessed Sep. 12, 2015, http://news.mit.edu/2015/identify-from-credit-card-metadata-0129.
3. Halliday, Josh, "Facebook users unwittingly revealing intimate secrets, study finds," *The Guardian*, Mar. 11, 2013, accessed Sep. 12, 2015, http://www.theguardian.com/technology/2013/mar/11/facebook-users-reveal-intimate-secrets.
4. Russell, Frank. *Information Gathering in Classical Greece*. Ann Arbor: University of Michigan Press, 1999.

Appendix A

Two Rights and a Wrong: Answers

Module 1

Difficulties in Preventing Attacks

1. Some attacks can vary their behavior so that the same attack appears differently.
2. Attack tools still require a high degree of skill and knowledge to use them.
3. The single difficulty that accounts for the greatest difficulty in preventing attacks is the increased speed of attacks.

Answer: The wrong statement is #2. Whereas at one time an attacker needed to have extensive technical knowledge of networks and computers as well as the ability to write a software program to generate an attack, that is no longer the case. Today's software attack tools do not require little if any sophisticated knowledge on the part of the attacker.

What Is Cybersecurity?

1. Cybersecurity is directly proportional to convenience.
2. Confidentiality ensures that only authorized parties can view the information.
3. Authentication ensures that the individual is who she claims to be and not an imposter.

Answer: The wrong statement is #1. The relationship between these two is not directly proportional (as security is increased, convenience is increased) but is inversely proportional (as security is increased, convenience is decreased). In other words, the more secure something becomes, the less convenient it may become to use.

Who Are the Attackers?

1. Script kiddies are responsible for the class of attacks called Advanced Persistent Threats.
2. Hactivists are strongly motivated by ideology.
3. Brokers sell their knowledge of a weakness to other attackers or a government.

Answer: The wrong statement is #1. State actors are responsible for the class of attacks called Advanced Persistent Threats.

Building a Comprehensive Security Strategy

1. Using layers can be a discouragement to the attackers to convince them to give up and find an easier target.
2. Updating defenses typically involves applying the latest updates sent from vendors to protect software and hardware.
3. There are four key elements to creating a practical security strategy.

   Answer: The wrong statement is #3. There are five key elements to creating a practical security strategy: block attacks, update defenses, minimize losses, use layers, and stay alert.

Module 2

Personal Security Attacks

1. Phishing is sending an email or displaying a web announcement that falsely claims to be from a legitimate enterprise in an attempt to trick the user into surrendering private information or taking action.
2. The most common IT authentication credential is providing information that only the user would know, namely a password.
3. The most effective password attack is a password spraying attack.

   Answer: The wrong statement is #3. A password spraying attack is a type of guessing at the user's password. A much more effective approach is to use password collections.

Creating a Defensive Stance

1. The key to protecting your digital life is to make it too difficult for an attacker to upend your safety, financial security, and privacy.
2. Being more secure than the average user will afford you the basic protections that you need.
3. No matter what defenses you use, they must be perfect to defeat all attackers.

   Answer: The wrong statement is #3. Defenses do not have to be perfect to be effective.

Personal Security Defenses

1. The most critical factor in a strong password is not length but complexity.
2. Password managers are not only used to store and retrieve passwords, but they also contain a password generator feature.
3. In a social engineering attack, the attacker presents herself as someone who can be trusted.

   Answer: The wrong statement is #1. The most critical factor in a strong password is not complexity but length: a longer password is always more secure than a shorter password. This is because the longer a password is, the more attempts an attacker must make to attempt to break it.

Module 3

Malware Attacks

1. The two types of viruses are a file-based virus and a fileless virus.
2. A keylogger can be a software program or a small hardware device.
3. When hundreds, thousands, or even millions of bot computers are gathered into a logical computer network, they create a "swarm."

Answer: The wrong statement is #3. When hundreds, thousands, or even millions of bot computers are gathered into a logical computer network, they create a botnet under the control of a bot herder.

Computer Defenses

1. Installing AV is the most important step to protecting your computer.
2. It is recommended that users have both a personal firewall and some type of hardware firewall.
3. The most comprehensive backup solution for most users is a continuous cloud backup.

Answer: The wrong statement is #1. Promptly installing patches once they are available is the most important step to protecting your computer.

Module 4

Internet Security Risks

1. When a website that uses JavaScript is accessed, the HTML document that contains the JavaScript code is downloaded onto the user's computer.
2. Extensions expand the normal capabilities of a web browser.
3. Spam, while annoying and a drain on productivity, is not considered dangerous.

Answer: The wrong statement is #3. One of the greatest risks of spam is that it is used to widely distribute malware.

Internet Defenses

1. Defending against Internet-based attacks begins with the foundation of first having the device itself properly secured.
2. HTTP is a secure protocol for sending information through the web.
3. Before installing a new extension, users should first check to see if this feature has already been added to the browser itself.

Answer: The wrong statement is #2. HTTPS is a secure protocol for sending information through the web.

Module 5

Mobile Attacks

1. A wireless router serves as a base station for the wireless devices, sending and receiving wireless signals between all devices as well as providing the access to the external Internet.
2. Bluetooth is a short-range wireless technology designed for the interconnection of two devices.
3. Downloading apps from an unofficial third-party app store is called jailbreaking.

Answer: The wrong statement is #3. Downloading apps from an unofficial third-party app store is called sideloading.

Mobile Defenses

1. The first step in securing a wireless router is to create a strong password to protect its internal configuration settings.
2. There is no known defense against connecting to an evil twin.
3. To prevent bluesnarfing, Bluetooth devices should be turned off when not being used or when in a room with unknown people.

Answer: The wrong statement is #2. One defense against connecting to an evil twin is to ask the establishment for the name of the official Wi-Fi network to prevent erroneously choosing an advertised evil twin network.

Module 6

Data Theft

1. Only about 10 percent of smartphone apps have tracking features.
2. As smartphones determine their current location, this information is then packaged into a feature on smartphones known as location services.
3. Surveillance-based advertising is targeted at an individual based on the characteristics of the individual.

Answer: The wrong statement is #1. Tracking features (*trackers*) are embedded in virtually every app on a smartphone.

Privacy Protections

1. A key is a mathematical value entered into the algorithm to produce the ciphertext.
2. Symmetric cryptography uses one key to encrypt data and a different key to decrypt data.
3. TLS protects messages only as they travel from the user's device to the app's servers, and then from the app's servers to the recipient's device.

Answer: The wrong statement is #2. Symmetric cryptography uses the same key to encrypt and decrypt the data.

Glossary

A

access point (AP) A more sophisticated device used in an office setting instead of a wireless router.

accounting The ability that provides tracking of events.

Advanced Persistent Threat (APT) A class of attacks by state actors that use innovative attack tools to silently extract data over an extended period.

antimalware software Software that can combat various malware attacks.

antivirus (AV) Antimalware software that can examine a computer for any infections as well as monitor computer activity and scan new documents that might contain a virus.

asset An item that has value.

asymmetric cryptography Cryptography that uses two keys.

attachment A document that is connected to an email message, such as a word processing document, spreadsheet, or picture.

attack vector The pathway for an attack.

authentication Proof of genuineness.

authorization The act of providing permission or approval to technology resources.

availability Security actions that ensure that data is accessible to authorized users.

B

backdoor Malware that gives access to a computer, program, or service that circumvents any normal security protections.

blocker ransomware Malware that prevents the user from using their computer in a normal fashion by blocking access to programs and files.

bluejacking An attack that sends unsolicited messages to Bluetooth-enabled devices.

bluesnarfing An attack that accesses unauthorized information from a wireless device through a Bluetooth connection.

Bluetooth A short-range wireless technology designed for quick interconnection of devices.

bot An infected computer placed under the remote control of an attacker to launch attacks.

broker An attacker who sells knowledge of a vulnerability to other attackers or governments.

C

code signing digital certificate A digital certificate used by software developers to digitally sign a program to prove that the software comes from the entity that signed it and no unauthorized third party has altered or compromised it.

command and control (C&C) A structure that sends instructions to infected bot computers.

Communications Assistance for Law Enforcement Act (CALEA) A wiretapping law passed in 1994 designed to enhance the ability of law enforcement agencies to conduct lawful interception of communications.

confidentiality Security actions that ensure that only authorized parties can view the information.

contactless payment system A system that uses near field communication (NFC) and provides an alternative to payment methods using cash or a credit card.

cookie User-specific information in a file stored on the user's computer and retrieved by a web server.

cross-site request forgery (CSRF) An attack that takes advantage of an authentication "token" that a website sends to a user's web browser to imitate the identity and privileges of the victim.

cross-site scripting (XSS) An attack that takes advantage of a website that accepts user input without validating it.

cryptography The practice of transforming information so that it cannot be understood by unauthorized parties and thus is secure.

cryptomalware Malware that encrypts files on a computer so that none of them can be opened until a ransom is paid.

cybercriminal An attacker whose goal is financial gain.

cybersecurity The tasks of protecting the integrity, confidentiality, and availability of information on the devices that store, manipulate, and transmit the information through products, people, and procedures.

cyberterrorism A politically motivated cyberattack designed to cause disruption and panic.

cyberterrorist An attacker whose motivation may be defined as ideological, or attacking for the sake of principles or beliefs.

D

data at rest Data that is stored on electronic media.

data backup A copy of the contents of a hard drive stored on other digital media in a secure location.

data in processing Data on which actions are being performed by devices.

data in transit Data on which actions are being performed as the data moves across a network.

decryption The cryptographic process of changing a scrambled message back into the original text.

digital certificate A technology used to associate a user's identity to a public key and that has been digitally signed by a trusted third party.

drive-by download An attack that results from a user visiting a specially crafted malicious webpage.

E

email digital certificate A digital certificate that allows a user to digitally sign and encrypt mail messages.

embedded hyperlink A link contained within the body of the message as a shortcut to a website.

encryption The cryptographic process of changing the original text into a scrambled message.

end-to-end encryption (E2EE) Encryption that protects messages in transit from the sender to the receiver.

evil twin An AP or another computer that is set up by an attacker designed to mimic the authorized Wi-Fi device.

extension A web browser addition that expands the normal capabilities of a web browser.

F

Fair and Accurate Credit Transactions Act (FACTA) of 2003 A law that contains rules regarding consumer privacy.

file-based virus Malicious computer code attached to a file so that each time the infected program is launched or the data file is opened, the virus unloads a payload and then reproduces itself.

fileless virus A type of virus that takes advantage of native services and processes to carry out its attacks.

firewall A packet filter designed to limit the spread of malware.

first-party cookie A cookie created from the website that a user is currently viewing.

full disk encryption (FDE) Cryptography that can be applied to entire disks instead of individual files or groups of files.

G

General Data Protection Regulation (GDPR) A directive that requires companies that perform business in the European Union (EU) to inform the EU's Information Commissioner's Office (ICO) if they suffer a breach involving the personal information of customers or employees.

geolocation The process of identifying the geographical location of the device.

GPS tagging (*geo-tagging*) Adding geographical identification data to media such as digital photos taken on a mobile device.

Gramm-Leach-Bliley Act (GLBA) A U.S. law that requires banks and financial institutions to alert customers of their policies and practices in disclosing customer information.

H

hacker An older term that referred to a person who used advanced computer skills to attack computers.

hactivist An attacker who attacks for ideological reasons.

Hardware Security Module (HSM) A removable external cryptographic hardware device.

Health Insurance Portability and Accountability Act (HIPAA) A U.S. law designed to guard protected health information and implement policies and procedures to safeguard it.

hoax A false warning, often contained in an email message claiming to come from the IT department.

hyperlink A link that connects one area on the web to another area that can be navigated with a click of the mouse button.

Hypertext Markup Language (HTML) A language that allows web authors to combine text, graphic images, audio, and video into a single document.

Hypertext Transfer Protocol (HTTP) A subset of a larger set of standards for Internet transmission.

Hypertext Transport Protocol Secure (HTTPS) A secure means of communication between a browser and a web server.

I

identity theft Stealing another person's personal information, such as a Social Security number, and then using the information to impersonate the victim, generally for financial gain.

impersonation Masquerading as a real or fictitious character and then playing the role of that person to trick a victim in a social engineering attack.

insider An employee, contractor, or business partner who is responsible for an attack.

Institute of Electrical and Electronics Engineers (IEEE) The most widely known and influential organization in the field of computer networking and wireless communications and sets wireless networking standards.

integrity Security actions that ensure that the information is correct and no unauthorized person or malicious software has altered the data.

Internet A global network that allows devices connected to it to exchange information.

Internet Mail Access Protocol (IMAP) A more recent and advanced email protocol.

J

jailbreaking Removing the built-in limitations and protections on Apple iOS devices.

JavaScript A popular scripting code that is embedded within HTML documents.

K

keylogger Hardware or software that silently captures and stores each keystroke a user types on the computer's keyboard.

L

location services A feature on smartphones that contains the current location of the device.

lock screen A technology that prevents a mobile device from being used until the user enters the correct passcode.

logic bomb Computer code that is typically added to a legitimate program but lies dormant and evades detection until a specific logical event triggers it.

M

macro A series of instructions that can be grouped as a single command.

Mail Transfer Agent (MTA) Programs that accept email messages from senders and route them toward their recipients.

Mail User Agent (MUA) Software used to read and send mail from a device.

malvertising Attacks based on malicious code sent through third-party advertising networks so that malware is distributed through ads sent to users' web browsers.

malware (*mal*icious soft*ware*) Software that enters a computer system without the user's knowledge or consent and then performs an unwanted and harmful action.

man-in-the-browser (MITB) An attack that intercepts communication between a browser and the underlying computer.

man-in-the-middle (MITM) An attack that intercepts legitimate communication to either eavesdrop on the conversation or impersonate one of the parties.

mesh A Wi-Fi network that consists of a main wireless router along with nodes.

mobile advertising identifier (MAID) A unique number that identifies a specific device.

multimedia messaging service (MMS) A technology that provides for pictures, video, or audio to be included in text messages.

N

near field communication (NFC) A set of standards used to establish communication between devices in very close proximity.

network firewall A hardware-based packet filter device that serves as the first line of defense for the network and devices connected to it.

O

online brute force attack An attack in which the same account is continuously attacked by entering different passwords.

P

password A secret combination of letters, numbers, and/or characters known only by the user.

password crackers Sophisticated attacker software that is specifically designed to break passwords.

password managers Technologies that are used to create, store, and retrieve passwords.

password spraying An attack in which one or a small number of commonly used passwords is used to attempt to log in to several different user accounts.

patch Software updates to address a security issue.

Payment Card Industry Data Security Standard (PCI DSS) A set of security standards that all U.S. companies processing, storing, or transmitting credit card information must follow.

personal firewall A software-based program that runs on the local computer to block or filter traffic coming into and out of the computer.

phishing Sending an email or displaying a web announcement that falsely claims to be from a legitimate enterprise in an attempt to trick the user into surrendering private information or taking action.

portable computer A device that is smaller than a desktop computer and is self-contained so that it can easily be transported while operating on battery power.

potentially unwanted program (PUP) Software that the user does not want on their computer.

PowerShell A task automation and configuration management framework from Microsoft.

private key An asymmetric cryptographic key that is known only to the individual to whom it belongs.

Protected View A Microsoft Office function that automatically opens selected documents in a read-only mode that disables editing functions.

public key An asymmetric cryptographic key that is known to everyone and can be freely distributed.

Q

Quick Response (QR) A two-dimensional barcode that consists of black modules arranged in a square grid on a white background.

R

random password generator A feature in password manager software to create long and unique passwords.

ransomware Malware that prevents a computer from properly and fully functioning until a fee is paid.

remote access Trojan (RAT) Malware that infects a computer like a Trojan but also gives the threat agent unauthorized remote access to the victim's computer.

rich communication services (RCS) A technology that can convert a texting app into a live chat platform and supports pictures, videos, location, stickers, and emojis.

risk A situation that involves exposure to danger.

rooting Removing the built-in limitations and protections on Google Android devices.

rootkit Malware that can hide its presence and the presence of other malware on the computer.

S

Sarbanes-Oxley Act (Sarbox) A U.S. law designed to fight corporate corruption.

script kiddie Individual who lacks advanced knowledge of computers and networks and so uses downloaded automated attack software.

self-encrypting drives (SEDs) Hard drives that can cryptographically protect all files stored on the device.

session cookie A cookie stored in random access memory (RAM), instead of on the hard drive, and only lasts for the duration of visiting the website.

session replay An attack in which an attacker attempts to impersonate the user by using the user's session token.

short message service (SMS) Text messages of a maximum of 160 characters.

sideloading Downloading an app from an unofficial third-party website.

smartphone A cellular phone with an operating system that allows it to run apps and access the Internet.

social engineering A means of using trickery to cause the victim to act in the attacker's favor.

social networking The use of Internet-based social media platforms that allow users to stay connected with friends, family, or peers.

spam Unsolicited email.

spam filter Software that inspects email messages to identify and stop spam.

spyware Tracking software that is deployed without the consent or control of the user.

state actor An attacker commissioned by a government.

surveillance-based advertising Internet-based digital advertising that is targeted at individuals who have been pre-identified through smartphone tracking data.

symmetric cryptography Cryptography that uses the same key to encrypt and decrypt data.

T

tablet A portable computing device that is generally larger than a smartphone and smaller than a laptop and is focused on ease of use.

third-party cookie A cookie from a website other than the original site.

threat A type of action that has the potential to cause harm.

threat actor A term used to describe individuals or entities who are responsible for cyber incidents against the technology equipment of enterprises and users.

threat agent A person or element that has the power to carry out a threat.

threat likelihood The probability that a threat will actually occur.

Transmission Control Protocol/Internet Protocol (TCP/IP) The standards for Internet transmissions.

Transport Layer Security (TLS) Encryption that protects messages only as they travel from the user's device to the app's servers, and then from the app's servers to the recipient's device.

Trojan An executable program that masquerades as performing a benign activity but also does something malicious.

Trusted Platform Module (TPM) A chip on the motherboard of the computer that provides cryptographic services.

two-factor authentication (2FA) Using two different credentials, such as something known and something in one's possession, for authentication.

typo squatting Registering fake websites for use when a victim misspells the web address of an actual site.

U

uniform resource locator (URL) A web address.

V

virtual private network (VPN) A technology that uses an unsecured public network, such as the Internet, as if it were a secure private network.

vulnerability A flaw or weakness that allows a threat agent to bypass security.

W

weak password A password that is easy for an attacker to break.

wearable A device that can be worn by the user instead of carried.

web browser Software that displays the words, pictures, and other elements following the instructions given through the HTML code.

Wi-Fi A wireless data network that provides high-speed data connections for mobile devices.

Wi-Fi Protected Access 2 (WPA2) Personal A security setting that provides strong security for all devices other than Wi-Fi 6E devices.

Wi-Fi Protected Access 3 (WPA3) Personal A security setting that provides strong security for Wi-Fi 6E devices.

Wi-Fi Protected Setup (WPS) A simplified and optional method for configuring WPA2 Personal wireless security that has security weaknesses.

wireless adapter A device that allows a mobile device to send and receive wireless signals.

wireless router A device used for a home-based Wi-Fi network that serves as the "base station" for wireless devices.

World Wide Web (WWW) A network composed of Internet server computers that provide online information in a specific format.

worm A malicious program that uses a computer network to replicate.

Index

A

accessing untrusted content in mobile devices, 126
access point (AP), 119
accounting, 7
Advanced Persistent Threat (APT), 16
algorithm
  cryptographic, 154
  hash, 30
allowed senders, 105
Android operating system, 125, 127
annual credit report, 52
antimalware software, 71–72
antivirus (AV) software, 71
  Defender Antivirus options, 72
Apple iOS, 125
application-based firewall, 73, 74
apps, 158
APT. *See* Advanced Persistent Threat (APT)
asset, 8
asymmetric cryptography, 157
ATM. *See* automated teller machine (ATM)
attachments, 104
  email, 92
attackers, 13–16
  brokers, 14
  characteristics of, 16
  cybercriminals, 13–14
  cyberterrorists, 15
  hacker, 13
  hactivists, 15
  insiders, 15
  script kiddies, 14
  state actors, 16
  types, 13
attacks
  authentication, 29
  brute force, 33
  cost of, 13
  difficulties in defending against, 2–5
  mobile, 116–127
  on mobile devices, 122–127
  online brute force attack, 33
  password spraying, 33
  on passwords, 33–34
  phishing, 35–37
  skills needed for creating, 14
  on wireless networks, 116–122
  tools menu, 3
  malware, 56–68
  using social engineering, 34–37
  vector, 9
  on Wi-Fi, 120
authentication, 7, 29
authorization, 7
automated teller machine (ATM), 133
automatic continuous backup, 75
availability, 6

B

backdoor, 68
block attacks, 17
blocked senders, 105
blocked top-level domain list, 105
blocker ransomware, 57
  computer infection, 59
  and malware, 58
  message, 58
bluejacking, 121
bluesnarfing, 121
Bluetooth, 120
  configuring, 132
  enabled devices, 121
  pairings, 121
  products, 120
  undiscoverable, 132
boomer barons, 152
bot, 67
bot herder, 67
botnets, 67
  attacks generated through, 67
  uses of, 67
brokers, 14
browser
  additions, 93, 94
  dangers, 92–94
  extensions, 94
  plug-ins, 94
  saving passwords in, 42
  scripting code, 93–94
brute force attack, 33

C

CALEA. *See* Communications Assistance for Law Enforcement Act (CALEA)
candidates, 33
card thieves, techniques of, 10
carriers, virus, 65
C&C or C2. *See* command and control (C&C or C2)
character set, 32
ciphertext, 154
cleartext data, 154. *See also* plaintext data
cloud backups, continuous, 75
cloud storage, 75
clusters, 152
code signing digital certificate, 161
command and control (C&C or C2), 67
Communications Assistance for Law Enforcement Act (CALEA), 150
computer defenses, 69–76
  creating data backups, 75–76
  examining firewalls, 72–74
  installing antimalware software, 71–72
  managing patches, 69–71
  stopping ransomware, 74–76
computer security, 55–86
computer virus. *See* virus
confidentiality, 6
connecting to public Wi-Fi, in mobile devices, 127
contactless payment systems, 121–122
continuous cloud backups, 75
convenience, relationship between security and, 5
cookies, 162
  session, 163
  types, 126
    first-party cookie, 163
    third-party cookie, 163
cross-site request forgery (CSRF), 99
cross-site scripting (XSS), 98–99
cryptography, 154–162
  algorithm of, 154
  and cleartext, 154
  cybersecurity benefits, 155–156
  and decryption, 154
  defined, 154
  and encryption, 154, 159
  information protections by, 156
  private key, 157
  process of, 155
  protections through, 159–162
  types of, 156–158
cryptomalware, 59
CSRF. *See* cross-site request forgery (CSRF)
cybercrime, 2, 13–14
cybercriminals, 13–14
cybersecurity, 5–13
  defining, 6–7
  importance, 9–12
    avoiding legal consequences, 11–12
    foil cyberterrorism, 11
    identity theft, 10
    maintaining productivity, 12
    preventing data theft, 9–10
  layers, 7–8
  and protection, 6–7
  terminology, 8–9
cyberterrorism foiling, 11
cyberterrorists, 15

D

data at rest, 156
data backups
  continuous cloud backups, 75
  creating, 75–76
  defined, 75
  scheduled local backup, 76
data breaches
  notifications, 23–24
  victim of, 24
  visual, 21–22
data brokers, 153
data in processing, 156
data in transit, 156, 160
data theft, 146–153
decryption, 154
delayed deletion, 75
digest, 30–31, 33
  message, 30–31, 33
digital certificates, 161
  code signing, 161
  email, 162
drive-by downloads, 98

E

E2EE. *See* end-to-end encryption (E2EE)
email, 90–92
  attachments, 92
  client, 105
  defenses, 104–106
  distributed malware, 95
  header, 91
  risks, 94–97

spam, 96
security settings, 105
web, 106
email digital certificate, 162
email MAU client, installed, 105
embedded hyperlinks, 95
encryption, 154
cryptographic hardware, 160
cryptographic software, 159–160
end-to-end encryption (E2EE), 162
evil twin, 127
network, 132
extensions, 94

F

Facebook, 38
likes indicated by users of, 152
FACTA. *See* Fair and Accurate Credit Transactions Act (FACTA) of 2003
factory settings, 134
Fair and Accurate Credit Transactions Act (FACTA) of 2003, 45
FDE. *See* full disk encryption (FDE)
file-based virus, 65
fileless virus, 64
firewalls, 72–74
application, 73
network, 74
personal, 73
Windows application, 73
first-party cookie, 163
fitness tracker, 124
flash cookie. *See* locally shared object (LSO)
full disk encryption (FDE), 160

G

gateway, 118
GDPR. *See* General Data Protection Regulation (GDPR)
General Data Protection Regulation (GDPR), 11
geolocation, 127
GLBA. *See* Gramm-Leach-Bliley Act (GLBA)
global positioning system (GPS), 148
Google Android, 124
GPS. *See* global positioning system (GPS)
GPS tagging, 127
Gramm-Leach-Bliley Act (GLBA), 12
graphical user interface (GUI), 69
guest access, turning on, 131
GUI. *See* graphical user interface (GUI)

H

hacker, 13
hactivists, 15
handoff, 119
hard disk drives, 60, 74, 75
Hardware Security Module (HSM), 160
hash algorithm, 30
Health Insurance Portability and Accountability Act (HIPAA), 11
HIPAA. *See* Health Insurance Portability and Accountability Act (HIPAA)
hoaxes, 37
home Wi-Fi security, 128–131
securing wireless router, 128–129
HTML. *See* Hypertext Markup Language (HTML)
HTML code, 89–90
HTTP. *See* Hypertext Transfer Protocol (HTTP)
HTTPS. *See* Hypertext Transport Protocol Secure (HTTPS)
hyperlinks, 89
Hypertext Markup Language (HTML), 89
Hypertext Transfer Protocol (HTTP), 90
Hypertext Transport Protocol Secure (HTTPS), 102

I

identity theft, 10, 38
avoiding, 44–45
IEEE. *See* Institute of Electrical and Electronics Engineers (IEEE)
image spam, 96
IMAP. *See* Internet Mail Access Protocol (IMAP)
impersonation, 35
information security. *See* cybersecurity
injecting malware, 98, 106, 120
insiders, 15
installing antimalware software, 71
installing unsecured applications, in mobile devices, 125
Institute of Electrical and Electronics Engineers (IEEE), 117
integrity, 6
Internet
defenses, 101–106
defined, 89
securing web browser, 90
security, 87–112
security risks, 92–97
browser dangers, 92–93
cookies, 162–163
cross-site request forgery, 99

Internet (*continued*)
cross-site scripting, 98–99
drive-by downloads, 98
extensions, 94
malvertising, 97–98
man-in-the-browser attack, 100
man-in-the-middle attack, 99–100
plug-in, 94
scripting code, 93–94
session replay attack, 100
tools, 89–92
email, 90–92
World Wide Web (WWW), 89–90
Internet Mail Access Protocol (IMAP), 91
Internet Protocol (IP) address, 128
IRS. *See* U.S. Internal Revenue Service (IRS)

J

jailbreaking, 125
JavaScript, 93

K

KeePass random password generator, 42
key, 154
keylogger, 61
and spyware, 62
kidnapping, 57

L

life stages, 152
limited physical security, in mobile devices, 126
living-off-the-land binaries (LOLBins), 66
local email client, 105
local security, 17
location services, 149
location tracking, in mobile devices, 127
lockout period, 134
lock screen, 132
logic bomb, 68
LOLBins. *See* living-off-the-land binaries (LOLBins)

M

macro, 94
MAID. *See* mobile advertising identifier (MAID)
Mail User Agent (MUA), 91
Mail Transfer Agent (MTA), 91
malicious attachments, 94–95
malvertising, 97–98
advantages for the attacker, 97–98
malware
attacks using, 56–68
blocker ransomware, 57
cryptomalware and, 59–60
defined, 56
injecting, 98, 106, 120
installing antimalware software, 71–72
keylogger, 61
kidnapping and, 57
and logic bomb, 68
masquerading, 62–68
backdoor, 68
bot, 67
logic bomb, 68
potentially unwanted programs, 63
remote access Trojan, 64
rootkit, 68
Trojan, 64
virus, 64–66
worm, 66–67
ransomware and, 57
spyware, 62
managing patches, 69–71
memorized password, 41
mesh, 119
message digest, 30
Microsoft Edge web browser security settings, 103, 111
Microsoft operating system updates, 69
Microsoft Windows security, configuring, 80–84
MITB. *See* man-in-the-browser (MITB) attack
MITM. *See* man-in-the-middle (MITM) attack
MMS. *See* multimedia messaging service (MMS)
mobile advertising identifier (MAID), 149, 163
mobile attacks, 116–127
mobile defenses, 128–135
mobile device security, 132–135
wireless network security, 128–132
mobile devices
attacks on, 122–127
best practices, 134
configuration, 132–134
disable unused features, 132–133
enable lock screen, 132
loss or theft, 134–135
security features for locating, 134–135
risks, 125–127
accessing untrusted content, 126
connecting to public Wi-Fi, 127
constrained updates, 127
installing unsecured applications, 125
limited physical security, 126
location tracking, 127

security, 132–135
types of, 123–125
portable computers, 124–125
smartphones, 124
tablets, 123
wearables, 124
man-in-the-browser (MITB) attack, 100
man-in-the-middle (MITM) attack, 100
mobile security, 115–143
MTA. *See* Mail Transfer Agent (MTA)
MUA. *See* Mail User Agent (MUA)
multimedia messaging service (MMS), 126

N

NAS. *See* network-attached storage (NAS) device
near field communication (NFC), 121
risks and defenses, 122
network-attached storage (NAS) device, 75
network firewall, 74
network of computer networks. *See* internet
network viruses. *See* worm
NFC. *See* near field communication (NFC)
nomophobia, 116

O

online backup services, 75
online brute force attack, 33
online or hardware-based restore, 75
online password cracker, 49–50
online vaults, password manager, 42
optional program file backup, 75

P

packet filter. *See* firewall
padlock icon and certificate information, 101–102
passphrase. *See* shared key
password, 29–34
on attacks, 29–31, 33
browser-based password management, 42
crackers and comparing password digests, 33–34
creating, 30
defenses, 40–41, 43
generators, 42
KeePass random password generator, 42
length, 43
managers, 41–43
memorized, 41
number of possible, 41
online cracker, 49–50
online vault password manager, 42
personal security and, 31–32
random password generator, 42
repeated, 41
retrieving, 31
spraying, 33
strong, 40–41
ten most common, 32
weaknesses, 31–32
patch, 69
Payment Card Industry Data Security Standard (PCI DSS), 12
PCI DSS. *See* Payment Card Industry Data Security Standard (PCI DSS)
personal firewall, 73
personal identification number (PIN), 130
personal security
attacks, 28–39
defenses, 40–46
password weaknesses, 31–32
phishing, 35–37
email message, 36
invoice scam, 36–37
PIN. *See* personal identification number (PIN)
plaintext data, 154. *See also* cleartext data
plug-in, 94
POP. *See* Post Office Protocol (POP)
POP3, 91
portable computers, 124
Post Office Protocol (POP), 91
potentially unwanted programs (PUPs), 63
PowerShell, 93–94
preshared key (PSK), 130
privacy, 145–165
best practices, 163–164
protections, 154–165
cryptography, 154–163
responsibilities of organizations, 165
private data
exfiltrated, 148–149
risks associated with, 152–153
associations with groups, 152
identity theft, 152
individual inconveniences, 152
statistical inferences, 152
unintended cross-pollination, 152
theft, 146–153
thieves, 149–151
types, 146–148, 151
private key, 157
cryptography, 157–158
productivity, maintaining, 12

protected view, 104
protection
  and accounting, 7
  and authentication, 7
  and authorization, 7
  and availability, 6
  and confidentiality, 6
  and information security, 7
  and integrity, 6
protocols, 90
public key, 157
public key cryptography. *See* asymmetric cryptographic algorithms
public Wi-Fi networks, 127
PUPs. *See* potentially unwanted programs (PUPs)

Q

QR codes. *See* quick response (QR) codes
quick response (QR) codes, 126
  creating and using, 141

R

radio frequency (RF) transmissions, 117
random password generator, 42. *See also* password
ransomware, 57
  blocker, 57–59
  stopping, 74–76
RAT. *See* remote access Trojan (RAT)
RCS. *See* rich communication services (RCS)
Reading pane, 105
remote access Trojan (RAT), 64
repeated password, 41
repudiation, 155
residential WLAN gateways, 118
rich communication services (RCS), 126
risk, 9. *See also* security risks
  email, 94–97
  mobile devices, 125–127
  near field communication (NFC), 122
  private data, 152–153
  social-networking, 37–38, 45–46
roaming, 119
rooting, 125
rootkit, 68
router
  configuring wireless, 142
  remote access settings, 129
  securing wireless, 128–129
routers, wireless. *See* wireless broadband routers

S

Sarbanes-Oxley Act (Sarbox), 11
scheduled local backups, 76
scripting code, 93–94
script kiddies, 14
secondary hard disk drive, 75
security
  comprehensive strategy of, 16–18
    block attacks, 17
    minimizing losses, 17
    staying alert, 18
    updating defenses, 17
    using layers, 17–18
  introduction to, 1–19
  local, 17
  patch, 69
  perimeter, 17
  relationship between convenience and, 5
  understanding, 5–6
SEDs. *See* self-encrypting drives (SEDs)
self-encrypting drives (SEDs), 160
Service Set Identifier (SSID), 131
session cookie, 163
session replay attack, 100
shared key, 130
short message service (SMS) text messages, 126
sideloading, 125
signature file, 71
Simple Mail Transfer Protocol (SMTP), 91
smartphones, 124
smartwatch, 124
SMTP. *See* Simple Mail Transfer Protocol (SMTP)
social engineering
  attacks using, 34–37, 43–44
  defenses, 44
  defined, 34
  effectiveness, 35
  hoax, 37
  identity theft, 38
  impersonation, 35
  phishing, 35–37
  typo squatting, 37
social-networking
  defined, 37
  risks, 37–38, 45–46
social security number, 38
spam, 96
  filters, 96–97, 105
  image, 96
spyware, 62
  and keylogger, 61
  technologies used by, 62

SSID. *See* Service Set Identifier (SSID)
state actors, 16
state-sponsored attackers, 16
statistical inferences, 152
strong password, 40–41
subnotebook, 125
surveillance-based advertising, 150
swipe pattern, 133
symmetric cryptography, 156–157

T

tablets, 123
TCP/IP. *See* Transmission Control Protocol/Internet Protocol (TCP/IP)
theft, identity, 10
third-party binary library, 94
third-party cookies, 163
threat, 8
  agent, 9
  likelihood, 9
  vector, 9
TLS. *See* Transport Layer Security (TLS)
Tomlinson, Ray, 90
TPM. *See* Trusted Platform Module (TPM)
Transmission Control Protocol/Internet Protocol (TCP/IP), 90
Transport Layer Security (TLS), 162
treasure-trove, 33
Trojan, 64
true blues, 152
trusted document, 104
trusted location, 104
Trusted Platform Module (TPM), 160
typo squatting, 37
two-factor authentication (2FA), 43

U

unblocking, 73
undiscoverable bluetooth, 132
uniform resource locator (URL), 90
universal access, 75
URL. *See* uniform resource locator (URL)
URL hijacking. *See* typo squatting
U.S. Internal Revenue Service (IRS), 10
USB storage device, unplugging, 75
username, 30

V

virtual private network (VPN), 132
virus, 64–66
  carriers, 65
  file-based, 65
  fileless, 66
  macro, 94
  vs. worms, 67
  Windows file types, infected with, 65
  Windows LOLBins, 66
VirusTotal, 84
VPN. *See* virtual private network (VPN)
vulnerability, 9

W

weak passwords, 31
wearables, 124
web. *See* World Wide Web (WWW)
web-based computer, 125
web browser, 90
  alternative, 112
  managing browser extensions, 104
  securing, 101
  security-related indicators, 101–102
  security settings, 102–104
  warnings, 102, 103
web email, 106
Wi-Fi equipment, 118–119
Wi-Fi networks, 117–120
  home, 119
  public, 127
  and Wi-Fi equipment, 118–119
Wi-Fi Protected Access 2 (WPA2) Personal, 130
  wireless router settings, 130
Wi-Fi Protected Access 3 (WPA3) Personal, 130–131
Wi-Fi Protected Setup (WPS), 130
Wi-Fi (wireless fidelity), 117
  attacks on, 120
  cells, 119
  names and standards, 117
  security settings, 131
Windows character map, 40
Windows 11, Microsoft
  operating system updates, 69
  patch update options, 70
  security settings on, 80–83
Windows personal firewall, 73
wireless adapter, 118
wireless broadband routers, 118
wireless local area network (WLAN), 117
wireless networks
  attacks on, 116
  and Bluetooth, 120–121
  near field communication, 121–122
  and Wi-Fi networks, 116–120
  security, 128–132

wireless routers, 118
  securing, 128–129
WLAN. *See* wireless local area network (WLAN)
World Wide Web (WWW), 89
worm, 66–67
  actions performed by, 66–67
  vs. virus, 67
WPS. *See* Wi-Fi Protected Setup (WPS)

XSS. *See* cross-site scripting (XSS)

zombie, 67